COMMON ERRORS
IN STATISTICS
(and How to
Avoid Them)

COMMON ERRORS IN STATISTICS (and How to Avoid Them)

Third Edition

P. I. GOOD AND J. W. HARDIN

A JOHN WILEY & SONS, INC., PUBLICATION

Published by John Wiley & Sons, Inc., Hoboken, New Jersey
Published simultaneously in Canada

For general information on our other products and services or for technical support, please contact our Customer Care Department within the United States at (800) 762-2974, outside the United States at (317) 572-3993 or fax (317) 572-4002.

Wiley also publishes its books in variety of electronic formats. Some content that appears in print may not be available in electronic format. For more information about Wiley products, visit our web site at www.wiley.com.

Library of Congress Cataloging-in-Publication Data:

Good, Phillip I.
  Common errors in statistics (and how to avoid them) / Phillip I. Good, James W. Hardin. -- 3rd ed.
    p. cm.
  Includes bibliographical references and index.
  ISBN 978-0-470-45798-6 (pbk.)
  1. Statistics. I. Hardin, James W. (James William) II. Title.
  QA276.G586 2009
  519.5--dc22

                                                                              2008047070

Printed in the United States of America

10 9 8 7 6 5 4 3 2

CONTENTS

PREFACE   **xi**

PART I   FOUNDATIONS   **1**

1   Sources of Error   **3**

Prescription, 4
Fundamental Concepts, 5
Ad Hoc, Post Hoc Hypotheses, 7
To Learn More, 11

2   Hypotheses: The Why of Your Research   **13**

Prescription, 13
What is a Hypothesis?, 14
Found Data, 16
Null Hypothesis, 16
Neyman–Pearson Theory, 17
Deduction and Induction, 21
Losses, 22
Decisions, 23
To Learn More, 25

3   Collecting Data   **27**

Preparation, 27
Response Variables, 28
Determining Sample Size, 32
Sequential Sampling, 36
One-Tail or Two?, 37
Fundamental Assumptions, 40
Experimental Design, 41
Four Guidelines, 43
Are Experiments Really Necessary?, 46
To Learn More, 47

PART II   STATISTICAL ANALYSIS   **49**

4   Data Quality Assessment   **51**

Objectives, 52
Review the Sampling Design, 52
Data Review, 53
The Four-Plot, 55
To Learn More, 55

5   Estimation   **57**

Prevention, 57
Desirable and Not-So-Desirable Estimators, 57
Interval Estimates, 61
Improved Results, 65
Summary, 66
To Learn More, 66

6   Testing Hypotheses: Choosing a Test Statistic   **67**

First Steps, 68
Test Assumptions, 70
Binomial Trials, 71
Categorical Data, 72
Time-to-Event Data (Survival Analysis), 73
Comparing the Means of Two Sets of Measurements, 76
Comparing Variances, 85
Comparing the Means of k Samples, 89
Subjective Data, 91
Independence Versus Correlation, 91
Higher-Order Experimental Designs, 92
Inferior Tests, 96

Multiple Tests, 97
Before You Draw Conclusions, 97
Summary, 99
To Learn More, 99

7   Miscellaneous Statistical Procedures   **101**

Bootstrap, 102
Bayesian Methodology, 103
Meta-Analysis, 110
Permutation Tests, 112
To Learn More, 113

PART III   REPORTS   **115**

8   Reporting Your Results   **117**

Fundamentals, 117
Descriptive Statistics, 122
Standard Error, 127
p-Values, 130
Confidence Intervals, 131
Recognizing and Reporting Biases, 133
Reporting Power, 135
Drawing Conclusions, 135
Summary, 136
To Learn More, 136

9   Interpreting Reports   **139**

With a Grain of Salt, 139
The Analysis, 141
Rates and Percentages, 145
Interpreting Computer Printouts, 146
To Learn More, 146

10   Graphics   **149**

The Soccer Data, 150
Five Rules for Avoiding Bad Graphics, 150
One Rule for Correct Usage of Three-Dimensional Graphics, 159
The Misunderstood and Maligned Pie Chart, 161
Two Rules for Effective Display of Subgroup Information, 162
Two Rules for Text Elements in Graphics, 166
Multidimensional Displays, 167
Choosing Graphical Displays, 170

Summary, 172
To Learn More, 172

PART IV   BUILDING A MODEL   **175**

11   Univariate Regression   **177**

Model Selection, 178
Stratification, 183
Estimating Coefficients, 185
Further Considerations, 187
Summary, 191
To Learn More, 192

12   Alternate Methods of Regression   **193**

Linear Versus Non-Linear Regression, 194
Least Absolute Deviation Regression, 194
Errors-in-Variables Regression, 196
Quantile Regression, 199
The Ecological Fallacy, 201
Nonsense Regression, 202
Summary, 202
To Learn More, 203

13   Multivariable Regression   **205**

Caveats, 205
Correcting for Confounding Variables, 207
Keep It Simple, 207
Dynamic Models, 208
Factor Analysis, 208
Reporting Your Results, 209
A Conjecture, 211
Decision Trees, 211
Building a Successful Model, 214
To Learn More, 215

14   Modeling Correlated Data   **217**

Common Sources of Error, 218
Panel Data, 218
Fixed- and Random-Effects Models, 219
Population-Averaged GEEs, 219
Quick Reference for Popular Panel Estimators, 221
To Learn More, 223

15   Validation   **225**

   Objectives, 225
   Methods of Validation, 226
   Measures of Predictive Success, 229
   Long-Term Stability, 231
   To Learn More, 231

GLOSSARY, GROUPED BY RELATED BUT DISTINCT TERMS   **233**

BIBLIOGRAPHY   **237**

AUTHOR INDEX   **259**

SUBJECT INDEX   **267**

PREFACE

One of the first times Dr. Good served as a statistical consultant, he was asked to analyze the occurrence rate of leukemia cases in Hiroshima, Japan, following World War II; on August 7, 1945, this city was the target site of the first atomic bomb dropped by the United States. Was the high incidence of leukemia cases among survivors the result of exposure to radiation from the atomic bomb? Was there a relationship between the number of leukemia cases and the number of survivors at certain distances from the bomb's epicenter?

To assist in the analysis, Dr. Good had an electric (not an electronic) calculator, reams of paper on which to write down intermediate results, and a prepublication copy of Scheffe's *Analysis of Variance*. The work took several months and the results were somewhat inclusive, mainly because he was never able to get the same answer twice—a consequence of errors in transcription rather than the absence of any actual relationship between radiation and leukemia.

Today, of course, we have high-speed computers and prepackaged statistical routines to perform the necessary calculations. Yet, statistical software will no more make one a statistician than a scalpel will turn one into a neurosurgeon. Allowing these tools to do our thinking is a sure recipe for disaster.

Pressed by management or the need for funding, too many research workers have no choice but to go forward with data analysis despite having insufficient statistical training. Alas, while a semester or two of undergraduate statistics may develop familiarity with the names of some statistical methods, it is not enough to make one aware of all the circumstances under which these methods may be applicable.

The purpose of this book is to provide a mathematically rigorous but readily understandable foundation for statistical procedures. Here are such basic concepts in statistics as null and alternative hypotheses, p-value, significance level, and power. Assisted by reprints from the statistical literature, we reexamine sample selection, linear regression, the analysis of variance, maximum likelihood, Bayes' Theorem, meta-analysis, and the bootstrap.

For the second and third editions, we've added material based on courses we've been offering at statcourse.com on unbalanced designs, report interpretation, and alternative modeling methods. The third edition has been enriched with even more examples from the literature. We've also added chapters on data quality assessment and general estimating equations.

More good news: Dr. Good's articles on women's sports have appeared in the *San Francisco Examiner*, *Sports Now*, and *Volleyball Monthly*, 22 short stories of his are in print, and you can find his novels on Amazon and Amazon/Kindle. So, if you can read the sports page, you'll find this book easy to read and to follow. Lest the statisticians among you believe that this book is too introductory, we point out the existence of hundreds of citations in statistical literature calling for the comprehensive treatment we have provided. Regardless of past training or current specialization, this book will serve as a useful reference; you will find applications for the information contained herein whether you are a practicing statistician or a well-trained scientist who just happens to apply statistics in the pursuit of other science.

The primary objective of Chapter 1 is to describe the main sources of error and provide a preliminary prescription for avoiding them. The hypothesis formulation – data gathering – hypothesis testing and estimate cycle is introduced, and the rationale for gathering additional data before attempting to test after-the-fact hypotheses is detailed.

A rewritten Chapter 2 places our work in the context of decision theory. We emphasize the importance of providing an interpretation of every potential outcome in advance data collection.

A much expanded Chapter 3 focuses on study design and data collection, as failure at the planning stage can render all further efforts valueless. The work of Berger and his colleagues on selection bias is given particular emphasis.

Chapter 4 on data quality assessment reminds us that just as 95% of research efforts are devoted to data collection, 95% of the time remaining should be spent on ensuring that the data collected warrant analysis.

Desirable features of point and interval estimates are detailed in Chapter 5 along with procedures for deriving estimates in a variety of practical situations. This chapter also serves to debunk several myths surrounding estimation procedures.

Chapter 6 reexamines the assumptions underlying testing hypotheses and presents the correct techniques for analyzing binomial trials, counts, categorical data, and continuous measurements. We review the impacts of violations of assumptions and detail the procedures to follow when making two- and k-sample comparisons.

Chapter 7 is devoted to the value and limitations of Bayes' Theorem, meta-analysis, the bootstrap, and permutation tests and contains essential tips on getting the most from these methods.

A much expanded Chapter 8 lists the essentials of any report that will utilize statistics, debunks the myth of the "standard" error, and describes the value and limitations of p-values and confidence intervals for reporting results. Practical significance is distinguished from statistical significance and induction is distinguished from deduction. Chapter 9 covers much the same material but from the viewpoint of the reader rather than the writer. Of particular importance is the section on interpreting computer output.

Twelve rules for more effective graphic presentations are given in Chapter 10 along with numerous examples of the right and wrong ways to maintain reader interest while communicating essential statistical information.

Chapters 11 through 15 are devoted to model building and to the assumptions and limitations of a multitude of regression methods and data mining techniques. A distinction is drawn between goodness of fit and prediction, and the importance of model validation is emphasized. Chapter 14 on modeling correlated data is entirely new. Seminal articles by David Freedman and Gail Gong are reprinted in the appendicies.

Finally, for the further convenience of readers, we provide a glossary grouped by related but contrasting terms, a bibliography, and subject and author indexes.

Our thanks to William Anderson, Leonardo Auslender, Vance Berger, Peter Bruce, Bernard Choi, Tony DuSoir, Cliff Lunneborg, Mona Hardin, Gunter Hartel, Fortunato Pesarin, Henrik Schmiediche, Marjorie Stinespring, and Peter A. Wright for their critical reviews of portions of this book. Doug Altman, Mark Hearnden, Elaine Hand, and David Parkhurst gave us a running start with their bibliographies. Brian Cade, David Rhodes, and the late Cliff Lunneborg helped us complete the second edition. Terry Therneau and Roswitha Blasche helped us complete the third edition.

We hope you soon put this book to practical use.

Sincerely yours,

Huntington Beach, CA   Phillip Good, drgood@statcourse.com
Columbia, SC   James Hardin, jhardin@gwm.sc.edu
December 2008

PART I

FOUNDATIONS

1

SOURCES OF ERROR

Don't think—use the computer.

—Dyke (tongue in cheek) [1997].

Statistical procedures for hypothesis testing, estimation, and model building are only a *part* of the decision-making process. They should never be used as the sole basis for making a decision (yes, even those procedures that are based on a solid deductive mathematical foundation). As philosophers have known for centuries, extrapolation from a sample or samples to a larger incompletely examined population must entail a leap of faith.

The sources of error in applying statistical procedures are legion and include all of the following:

- Using the same set of data to formulate hypotheses and then to test those hypotheses.
- Taking samples from the wrong population or failing to specify in advance the population(s) about which inferences are to be made.
- Failing to draw samples that are random and representative.
- Measuring the wrong variables or failing to measure what you intended to measure.
- Failing to understand that *p*-values are statistics, that is, functions of the observations, and will vary in magnitude from sample to sample.
- Using inappropriate or inefficient statistical methods.
- Using statistical software without verifying that its current defaults are appropriate for your application.
- Failing to validate models.

Common Errors in Statistics (and How to Avoid Them), Third Edition. Edited by P. I. Good and J. W. Hardin
Copyright © 2009 John Wiley & Sons, Inc.

But perhaps the most serious source of error is letting statistical procedures make decisions for you.

In this chapter, as throughout this book, we first offer a preventive prescription, followed by a list of common errors. If these prescriptions are followed carefully, you will be guided to the correct and effective use of statistics and avoid the pitfalls.

PRESCRIPTION

Statistical methods used for experimental design and analysis should be viewed in their rightful role as merely a part, albeit an essential part, of the decision-making procedure.

Here is a partial prescription for the error-free application of statistics:

1. Set forth your objectives and your research intentions *before* you conduct a laboratory experiment, a clinical trial, or a survey or analyze an existing set of data.

2. Define the population about which you will make inferences from the data you gather.

3. List all possible sources of variation. Control them or measure them to avoid confounding them with relationships among those items that are of primary interest.

4. Formulate your hypotheses and all of the associated alternatives. (See Chapter 2.) List possible experimental findings along with the conclusions you would draw and the actions you would take if this or another result proves to be the case. Do all of these things *before* you complete a single data collection form and *before* you turn on your computer.

5. Describe in detail how you intend to draw a representative sample from the population. (See Chapter 3.)

6. Use estimators that are impartial, consistent, efficient, robust, and minimum loss. (See Chapter 5.) To improve the results, focus on sufficient statistics, pivotal statistics, and admissible statistics and use interval estimates. (See Chapters 5 and 6.)

7. Know the assumptions that underlie the tests you use. Use those tests that require the minimum number of assumptions and are most powerful against the alternatives of interest. (See Chapters 5, 6, and 7.)

8. Incorporate in your reports the complete details of how the sample was drawn and describe the population from which it was drawn. If data are missing or the sampling plan was not followed, explain why and list all differences between the data that were present in the sample and the data that were missing or excluded. (See Chapter 8.)

FUNDAMENTAL CONCEPTS

Three concepts are fundamental to the design of experiments and surveys: variation, population, and sample.

A thorough understanding of these concepts will forestall many errors in the collection and interpretation of data.

> If there were no variation—if every observation were predictable, a mere repetition of what had gone before—there would be no need for statistics.

Variation

Variation is inherent in virtually all of our observations. We would not expect the outcomes of two consecutive spins of a roulette wheel to be identical. One result might be red, the other black. The outcome varies from spin to spin.

There are gamblers who watch and record the spins of a single roulette wheel hour after hour, hoping to discern a pattern. A roulette wheel is, after all, a mechanical device, and perhaps a pattern will emerge. But even those observers do not anticipate finding a pattern that is 100% predetermined. The outcomes are just too variable.

Anyone who spends time in a schoolroom, as a parent or as a child, can see the vast differences among individuals. This one is tall, that one is short, though all are the same age. Half an aspirin and Dr. Good's headache is gone, but his wife requires four times that dosage.

There is variability even among observations on deterministic formula-satisfying phenomena such as the position of a planet in space or the volume of gas at a given temperature and pressure. Position and volume satisfy Kepler's laws and Boyle's law, respectively, but the observations we collect will depend upon the measuring instrument (which may be affected by the surrounding environment) and the observer. Cut a length of string and measure it three times. Do you record the same length each time?

In designing an experiment or a survey, we must always consider the possibility of errors arising from the measuring instrument and from the observer. It is one of the wonders of science that Kepler was able to formulate his laws at all given the relatively crude instruments at his disposal.

Population

> The population(s) of interest must be clearly defined before we begin to gather data.

From time to time, someone will ask us how to generate confidence intervals (see Chapter 8) for the statistics arising from a total census of a population. Our answer is that we cannot help. Population statistics (mean, median, 30th percentile) are not estimates. They are fixed values and will be known with 100% accuracy if two criteria are fulfilled:

1. Every member of the population is observed.
2. All the observations are recorded correctly.

Confidence intervals would be appropriate if the first criterion is violated, for then we are looking at a sample, not a population. And if the second criterion is violated, then we might want to talk about the confidence we have in our measurements.

Debates about the accuracy of the 2000 United States Census arose from doubts about the fulfillment of these criteria.[1] "You didn't count the homeless" was one challenge. "You didn't verify the answers" was another. Whether we collect data for a sample or an entire population, both of these challenges or their equivalents can and should be made.

Kepler's "laws" of planetary movement are not testable by statistical means when applied to the original planets (Jupiter, Mars, Mercury, and Venus) for which they were formulated. But when we make statements such as "Planets that revolve around Alpha Centauri will also follow Kepler's laws," we begin to view our original population, the planets of our sun, as a sample of all possible planets in all possible solar systems.

A major problem with many studies is that the population of interest is not adequately defined before the sample is drawn. Don't make this mistake. A second major problem is that the sample proves to have been drawn from a different population than was originally envisioned. We consider these issues in the next section and again in Chapters 2, 6, and 7.

Sample

A sample is any (proper) subset of a population.

Small samples may give a distorted view of the population. For example, if a minority group comprises 10% or less of a population, a jury of 12 persons selected at random from that population fails to contain any members of that minority at least 28% of the time.

As a sample grows larger, or as we combine more clusters within a single sample, the sample will resemble more closely the population from which it is drawn.

How large a sample must be drawn to obtain a sufficient degree of closeness will depend upon the manner in which the sample is chosen from the population.

Are the elements of the sample drawn at random, so that each unit in the population has an equal probability of being selected? Are the elements of the sample drawn independently of one another? If either of these criteria is not satisfied, then even a very large sample may bear little or no relation to the population from which it was drawn.

---

[1] *City of New York v. Department of Commerce*, 822 F. Supp. 906 (E.D.N.Y, 1993). The arguments of four statistical experts who testified in the case may be found in Volume 34 of *Jurimetrics*, 1993, 64–115.

An obvious example is the use of recruits from a Marine boot camp as representatives of the population as a whole or even as representatives of all Marines. In fact, any group or cluster of individuals who live, work, study, or pray together may fail to be representative for any or all of the following reasons [Cummings and Koepsell, 2002]:

1. Shared exposure to the same physical or social environment.
2. Self-selection in belonging to the group.
3. Sharing of behaviors, ideas, or diseases among members of the group.

A sample consisting of the first few animals to be removed from a cage will not satisfy these criteria either, because, depending on how we grab, we are more likely to select more active or more passive animals. Activity tends to be associated with higher levels of corticosteroids, and corticosteroids are associated with virtually every body function.

Sample bias is a danger in every research field. For example, Bothun [1998] documents the many factors that can bias sample selection in astronomical research.

To forestall sample bias in your studies, before you begin, determine all the factors that can affect the study outcome (gender and lifestyle, for example). Subdivide the population into strata (males, females, city dwellers, farmers) and then draw separate samples from each stratum. Ideally, you would assign a random number to each member of the stratum and let a computer's random number generator determine which members are to be included in the sample.

Surveys and Long-term Studies

Being selected at random does not mean that an individual will be willing to participate in a public opinion poll or some other survey. But if survey results are to be representative of the population at large, then pollsters must find some way to interview nonresponders as well. This difficulty is exacerbated in long-term studies, as subjects fail to return for follow-up appointments and move without leaving a forwarding address. Again, if the sample results are to be representative, some way must be found to report on subsamples of the nonresponders and the dropouts.

AD HOC, POST HOC HYPOTHESES

Formulate and write down your hypotheses before you examine the data.

Patterns in data can suggest but cannot confirm hypotheses unless these hypotheses were formulated *before* the data were collected.

Everywhere we look, there are patterns. In fact, the harder we look, the more patterns we see. Three rock stars die in a given year. Fold the U.S. $20 bill in just the right way, and not only the Pentagon but also the Twin Towers in flames are revealed.[2] It is natural for us to want to attribute some underlying cause to these patterns. But those who have studied the laws of probability tell us that more often than not, patterns are simply the result of random events.

Put another way, there is a greater probability of finding at least one cluster of events in time or space than finding no clusters at all (equally spaced events).

How can we determine whether an observed association represents an underlying cause-and-effect relationship or is merely the result of chance? The answer lies in our research protocol. When we set out to test a specific hypothesis, the probability of a specific event is predetermined. But when we uncover an apparent association, one that may have arisen purely by chance, we cannot be sure of the association's validity until we conduct a second set of controlled trials.

In the International Study of Infarct Survival [1988], patients born under the Gemini or Libra astrological birth signs did not survive as long when their treatment included aspirin. By contrast, aspirin offered an apparent beneficial effect (longer survival time) to study participants with all other astrological birth signs.

Except for those who guide their lives by the stars, there is no hidden meaning or conspiracy in this result. When we describe a test as significant at the 5% or 1-in-20 level, we mean that 1 in 20 times we'll get a significant result even though the hypothesis is true. That is, when we test to see if there are any differences in the baseline values of the control and treatment groups, if we've made 20 different measurements, we can expect to see at least one statistically significant difference; in fact, we will see this result almost two-thirds of the time. This difference will not represent a flaw in our design but simply chance at work. To avoid this undesirable result—that is, to avoid attributing statistical significance to an insignificant random event, a so-called Type I error—we must distinguish between the hypotheses with which we began the study and those which came to mind afterward. We must accept or reject these hypotheses at the original significance level while demanding additional corroborating evidence for those exceptional results (such as dependence of an outcome on an astrological sign) that are uncovered for the first time during the trials.

No reputable scientist would ever report results before successfully reproducing the experimental findings twice, once in the original laboratory and once in that of a colleague.[3] The latter experiment can be particularly telling, as all too often some overlooked factor not controlled in the experiment—such as the quality of the laboratory water—proves responsible for the results observed initially. It's better to be found wrong in private than in public. The only remedy is to attempt to replicate the findings with different sets of subjects, replicate, then replicate again.

Persi Diaconis [1978] spent some years investigating paranormal phenomena. His scientific inquiries included investigating the powers linked to Uri Geller (Fig. 1.1), the man who claimed he could bend spoons with his mind. Diaconis was not surprised

---

[2]A website with pictures is located at http://www.foldmoney.com/.

[3]Remember "cold fusion?" In 1989, two University of Utah professors told the newspapers that they could fuse deuterium molecules in the laboratory, solving the world's energy problems for years to come. Alas, neither those professors nor anyone else could replicate their findings, though true believers abound; see http://www.ncas.org/erab/intro.htm.

Figure 1.1. Photo of Uri Geller. *Source*: Reprinted with permission of Aquarius 2000 of the German Language Wikipedia.

to find that the hidden "powers" of Geller were more or less those of the average nightclub magician, down to and including forcing a card and taking advantage of ad hoc, post hoc hypotheses.

When three buses show up at your stop simultaneously, or three rock stars die in the same year, or a stand of cherry trees is found amid a forest of oaks, a good statistician remembers the Poisson distribution. This distribution applies to relatively rare events that occur independently of one another. The calculations performed by Siméon-Denis Poisson reveal that if there is an average of one event per interval (in time or in space), then while more than a third of the intervals will be empty, at least a quarter of the intervals are likely to include multiple events (Fig. 1.2).

Anyone who has played poker will concede that one out of every two hands contains something interesting (Table 1.1). Don't allow naturally occurring results to fool you or lead you to fool others by shouting "Isn't this incredible?"

The purpose of a recent set of clinical trials was to see if blood flow and distribution in the lower leg could be improved by carrying out a simple surgical procedure prior to the administration of standard prescription medicine.

The results were disappointing on the whole, but one of the marketing representatives noted that the long-term prognosis was excellent when a marked increase in blood flow was observed just after surgery. She suggested that we calculate a

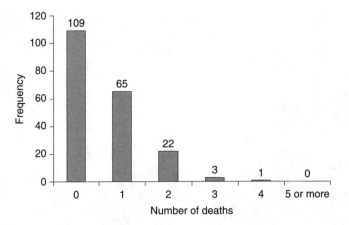

Figure 1.2. Frequency plot of the number of deaths in the Prussian army as a result of being kicked by a horse (200 total observations).

TABLE 1.1. Probability of Finding Something Interesting in a Five-Card Hand

Hand	Probability
Straight flush	0.0000
Four of a kind	0.0002
Full house	0.0014
Flush	0.0020
Straight	0.0039
Three of a kind	0.0211
Two pairs	0.0475
Pair	0.4226
Total	0.4988

p-value[4] for a comparison of patients with improved blood flow after surgery versus patients who had taken the prescription medicine alone.

Such a p-value is meaningless. Only one of the two samples of patients in question had been taken at random from the population (those patients who received the prescription medicine alone). The other sample (those patients who had increased blood flow following surgery) was determined after the fact. To extrapolate results from the samples in hand to a larger population, the samples must be taken at random from, and be representative of, that population.

The preliminary findings clearly called for an examination of surgical procedures and of patient characteristics that might help forecast successful surgery. But the

---

[4] A p-value is the probability under the primary hypothesis of observing the set of observations we have in hand. We can calculate a p-value once we make a series of assumptions about how the data were gathered. Today statistical software does the calculations, but it's still up to us to validate the assumptions.

generation of a p-value and the drawing of any final conclusions had to wait on clinical trials specifically designed for that purpose.

This doesn't mean that one should not report anomalies and other unexpected findings. Rather, one should not attempt to provide p-values or confidence intervals in support of them. Successful researchers engage in a cycle of theorizing and experimentation so that the results of one experiment become the basis for the hypotheses tested in the next.

A related, extremely common error whose correction we discuss at length in Chapters 13 and 14 is to use the same data to select variables for inclusion in a model and to assess their significance. Successful model builders develop their frameworks in a series of stages, validating each model against a second independent data set before drawing conclusions.

TO LEARN MORE

On the necessity for improvements in the use of statistics in research publications, see Altman [1982, 1991, 1994, 2000, 2002], Cooper and Rosenthal [1980], Dar, Serlin, and Omer [1994], Gardner and Bond [1990], George [1985], Glantz [1980], Goodman, Altman, and George [1998], MacArthur and Jackson [1984], Morris [1988], Thorn, Pilliam, Symons, and Eckel [1985], and Tyson, Furzan, Reisch and Mize [1983].

2

HYPOTHESES: THE WHY OF YOUR RESEARCH

In this chapter, we review how to formulate a hypothesis that is testable by statistical means, the appropriate use of the null hypothesis, Neyman–Pearson theory, the two types of error, and the more general theory of decisions and losses.

PRESCRIPTION

Statistical methods used for experimental design and analysis should be viewed in their rightful role as merely a part, albeit an essential part, of the decision-making procedure.

1. Set forth your objectives and the use you plan to make of your research *before* you conduct a laboratory experiment, clinical trial, or survey or analyze an existing set of data.

2. Formulate your hypothesis and *all* of the associated alternatives. List possible experimental findings along with the conclusions you would draw and the actions you would take if this or another result proves to be the case. Do all of these things *before* you complete a single data collection form and *before* you turn on your computer.

Common Errors in Statistics (and How to Avoid Them), Third Edition. Edited by P. I. Good and J. W. Hardin
Copyright © 2009 John Wiley & Sons, Inc.

WHAT IS A HYPOTHESIS?

A well-formulated hypothesis will be both quantifiable and testable, that is, involve measurable quantities or refer to items that may be assigned to *mutually exclusive* categories.

A well-formulated statistical hypothesis takes one of two forms: "Some measurable characteristic of a population takes one of a specific set of values" or "Some measurable characteristic takes different values in different populations, the difference(s) having a specific pattern or a specific set of values."

Examples of well-formed statistical hypotheses include the following:

- "For males over 40 suffering from chronic hypertension, a 100 mg daily dose of this new drug will lower diastolic blood pressure an average of 10 mm Hg."
- "For males over 40 suffering from chronic hypertension, a daily dose of 100 mg of this new drug will lower diastolic blood pressure an average of 10 mm Hg more than an equivalent dose of metoprolol."
- "Given less than 2 hours per day of sunlight, applying 1 to 10 lb of 23-2-4 fertilizer per 1000 ft^2 will have no effect on the growth of fescues and Bermuda grasses."

"All redheads are passionate" is not a well-formed statistical hypothesis, not merely because "passionate" is not defined, but also because the word "All" suggests that there is no variability. The latter problem can be solved by quantifying the term "All" to perhaps 80%. If we specify "passionate" in quantitative terms to mean "has an orgasm more than 95% of the time consensual sex is performed," then the hypothesis "80% of redheads have an orgasm more than 95% of the time consensual sex is performed" becomes testable.

Note that defining "passionate" to mean "has an orgasm *every time* consensual sex is performed" would not be provable, as it too is a statement of the "all or none" variety. The same is true for a hypothesis like "has an orgasm *none* of the times consensual sex is performed." Similarly, qualitative assertions of the form "Not all" or "Some" are not statistical in nature because these terms leave much room for subjective interpretation. How many do we mean by "some"? Five out of 100? Ten out of 100?

The statements "Doris J. is passionate" and "Both Good brothers are 5'10" tall" are also not statistical in nature, as they concern specific individuals rather than populations [Hagood, 1941]. Finally, note that until someone other than Thurber succeeds in locating unicorns, the hypothesis, "80% of unicorns are white" is *not* testable.

Formulate your hypotheses so that they are quantifiable, testable, and statistical in nature.

How Precise Must a Hypothesis Be?

The chief executive of a drug company may well express a desire to test whether "our antihypertensive drug can beat the competition." The researcher might want to test a preliminary hypothesis on the order of "For males over 40 suffering from chronic hypertension, there is a daily dose of our new drug which will lower diastolic blood pressure an average of 20 mm Hg." But this hypothesis is imprecise. What if getting the necessary dose of the new drug required taking a tablet every hour? Or caused liver malfunction or even death? First, the researcher would need to conduct a set of clinical trials to determine the maximum tolerable dose (MTD). Subsequently, she could test the precise hypothesis "A daily dose of one-third to one-fourth the MTD of our new drug will lower diastolic blood pressure an average of 20 mm Hg in males over 40 suffering from chronic hypertension."

A BILL OF RIGHTS

- Scientists can and should be encouraged to make subgroup analyses.
- Physicians and engineers should be encouraged to make decisions utilizing the findings of such analyses.
- Statisticians and other data analysts can and should refuse to give their imprimatur to related tests of significance.

In a series of articles by Horwitz et al. [1998], a physician and his colleagues strongly criticize the statistical community for denying them (or so they perceive) the right to provide a statistical analysis for subgroups not contemplated in the original study protocol. For example, suppose that in a study of the health of Marine recruits, we notice that not one of the dozen or so women who received a vaccine contracted pneumonia. Are we free to provide a *p*-value for this result?

Statisticians Smith and Egger [1998] argue against hypothesis tests of subgroups chosen after the fact, suggesting that the results are often likely to be explained by the "play of chance." Altman [1998b, pp. 301–303], another statistician, concurs noting that "... the observed treatment effect is expected to vary across subgroups of the data ... simply through chance variation" and that "doctors seem able to find a biologically plausible explanation for any finding." This leads Horwitz et al. to the incorrect conclusion that Altman proposes that we "dispense with clinical biology (biologic evidence and pathophysiologic reasoning) as a basis for forming subgroups." Neither Altman nor any other statistician would quarrel with Horwitz et al.'s assertion that physicians must investigate "how ... we [physicians] do our best for a particular patient."

Scientists can and should be encouraged to make subgroup analyses. Physicians and engineers should be encouraged to make decisions based upon them. Few would deny that in an emergency, coming up with workable, fast-acting solutions without complete information is better than *finding the best possible solution*.[1] But, by the same token, statisticians should not be pressured to give their imprimatur to

---

[1]Chiles [2001, p. 61].

what, in statistical terms, is clearly an improper procedure, nor should statisticians mislabel suboptimal procedures as the best that can be done.[2]

We concur with Anscombe [1963], who writes, "...the concept of error probabilities of the first and second kinds...has no direct relevance to experimentation.... The formation of opinions, decisions concerning further experimentation and other required actions, are not dictated...by the formal analysis of the experiment, but call for judgment and imagination.... It is unwise for the experimenter to view himself seriously as a decision-maker.... The experimenter pays the piper and calls the tune he likes best; but the music is broadcast so that others might listen...."

FOUND DATA

p-Values should not be computed for hypotheses based on "found data," as of necessity, all hypotheses related to found data are formulated after the fact. This rule does not apply if the observer first divides the data into sections. One part is studied and conclusions are drawn; then the resultant hypotheses are tested on the remaining sections. Even then, the tests are valid only if the found data can be shown to be representative of the population at large.

NULL HYPOTHESIS

> A major research failing seems to be the exploration of uninteresting or even trivial questions.... In the 347 sampled articles in *Ecology* containing null hypotheses tests, we found few examples of null hypotheses that seemed biologically plausible.
> —Anderson, Burnham, and Thompson [2000]

> We do not perform an experiment to find out if two varieties of wheat or two drugs are equal. We know in advance, without spending a dollar on an experiment, that they are not equal.
> —Deming [1975]

> Test only relevant null hypotheses.

The null hypothesis has taken on an almost mythic role in contemporary statistics. Obsession with the null hypothesis has been allowed to shape the direction of our research. We've let the tool use us instead of our using the tool.[3]

While a null hypothesis can facilitate statistical inquiry—an exact permutation test (as discussed in Chapters 5 and 7) is impossible without it—it is never mandated. In

---

[2]One is reminded of the dean—several of them, in fact—who asked me to alter my grades.

  "But that is something, you can do as easily as I."

  "Why, Dr. Good, I would never dream of overruling one of my instructors."

[3]See, for example, Hertwig and Todd [2000].

any event, virtually any quantifiable hypothesis can be converted into null form. There is no excuse and no need to be content with a meaningless null.

For example, suppose we want to test that a given treatment will decrease the need for bed rest by at least three days. Previous trials have convinced us that the treatment will reduce the need for bed rest to some degree, so merely testing that the treatment has a positive effect would yield no new information. Instead, we would subtract three from each observation and then test the null hypothesis that the mean value is zero.

We often will want to test that an effect is inconsequential, not zero but close to it, smaller than d, say, where d is the smallest biological, medical, physical, or socially relevant effect in our area of research. Again, we would subtract d from each observation before proceeding to test a null hypothesis.

The quote from Deming above is not quite correct, as often we will wish to demonstrate that two drugs or two methods yield equivalent results. As shown in Chapter 5, we test for equivalence using confidence intervals; a null hypothesis is not involved.

To test that "80% of redheads are passionate," we have two choices, depending on how passion is measured. If passion is an all-or-none phenomenon, then we can forget about trying to formulate a null hypothesis and instead test the binomial hypothesis that the probability p that a redhead is passionate is 80%. If passion can be measured on a 7-point scale and we define "passionate" as passion greater than or equal to 5, then our hypothesis becomes "The 20th percentile of redhead passion exceeds 5." As in the first example above, we could convert this to a null hypothesis by subtracting five from each observation. But the effort is unnecessary, as this problem, too, reduces to a test of a binomial parameter.

NEYMAN–PEARSON THEORY

Formulate your alternative hypotheses at the same time you set forth the hypothesis that is of chief concern to you.

When the objective of our investigations is to arrive at some sort of conclusion, we must have not only a single primary hypothesis in mind but also one or more potential alternative hypotheses.

The cornerstone of modern hypothesis testing is the Neyman–Pearson Lemma. To get a feeling for the working of this mathematical principle, suppose we are testing a new vaccine by administering it to half of our test subjects and giving a supposedly harmless placebo to the remainder. We proceed to follow these subjects over some fixed period and note which subjects, if any, contract the disease that the new vaccine is said to offer protection against.

We know in advance that the vaccine is unlikely to offer complete protection; indeed, some individuals may actually come down with the disease as a result of taking the vaccine. Many factors over which we have no control, such as the weather, may result in none of the subjects, even those who received only placebo, contracting the disease during the study period. All sorts of outcomes are possible.

The tests are being conducted in accordance with regulatory agency guidelines. From the regulatory agency's perspective, the principal hypothesis, H, is that the new vaccine offers no protection. The alternative hypothesis, A, is that the new vaccine can cut the number of infected individuals in half. Our task before the start of the experiment is to decide which outcomes will rule in favor of the alternative hypothesis A and which in favor of the null hypothesis H.

The problem is that because of the variation inherent in the disease process, every one of the possible outcomes could occur regardless of which hypothesis is true. Of course, some outcomes are more likely if H is true—for example, 50 cases of pneumonia in the placebo group and 48 in the vaccine group—and others are more likely if the alternative hypothesis is true—for example, 38 cases of pneumonia in the placebo group and 20 in the vaccine group.

Following Neyman and Pearson, we order each of the possible outcomes in accordance with the ratio of its probability or likelihood when the alternative hypothesis is true to its probability when the principal hypothesis is true. When this likelihood ratio is large, we say the outcome rules in favor of the alternative hypothesis. Working downward from the outcomes with the highest values, we continue to add outcomes to the *rejection* region of the test—so called because these are the outcomes for which we would reject the primary hypothesis—until the total probability of the rejection region under the primary hypothesis is equal to some predesignated *significance level*.

In the following example, we would reject the primary hypothesis at the 10% level only if the test subject really liked a product.

	Really Hate	Dislike	Indifferent	Like	Really Like
Primary hypothesis	10%	20%	40%	20%	10%
Alternate hypothesis	5%	10%	30%	30%	25%
Likelihood ratio	1/2	1/2	3/4	3/2	5/2

To see that we have done the best we can do, suppose we replace one of the outcomes we assigned to the rejection region with one we did not. The probability that this new outcome would occur if the primary hypothesis is true must be less than or equal to the probability that the outcome it replaced would occur if the primary hypothesis is true. Otherwise, we would exceed the significance level. Because of the way we assigned outcome to the rejection region, the likelihood ratio of the new outcome is smaller than the likelihood ratio of the old outcome. Thus, the probability that the new outcome would occur if the alternative hypothesis is true must be less than or equal to the probability that the outcome it replaced would occur if the alternative hypothesis is true. That is, by swapping outcomes we have reduced the *power* of our test. By following the method of Neyman and Pearson and maximizing the likelihood ratio, we obtain the most powerful test at a given significance level.

To take advantage of Neyman and Pearson's finding, we need to have one or more alternative hypotheses firmly in mind when we set up a test. Too often in published research, such alternative hypotheses remain unspecified or, worse, are specified only *after* the data are in hand. *We must specify our alternatives before we commence an analysis*, preferably at the same time we design our study.

Are our alternatives one-sided or two-sided? If we are comparing several populations at the same time, are there means ordered or unordered? The form of the alternative will determine the statistical procedures we use and the significance levels we obtain.

> Decide beforehand whether you wish to test against a one-sided or a two-sided alternative.

One-sided or Two-sided

Suppose that, on examining the cancer registry in a hospital, we uncover the following data that we put in the form of a 2×2 contingency table:

	Survived	Died	Total
Men	9	1	10
Women	4	10	14
Total	13	11	24

The 9 denotes the number of men who survived, the 1 denotes the number of men who died, and so forth. The four marginal totals, or marginals, are 10, 14, 13, and 11. The total number of men in the study is 10, while 14 denotes the total number of women, and so forth.

The marginals in this table are fixed because, indisputably, there are 11 dead bodies among the 24 persons in the study, including 14 women. Suppose that before completing the table, we lost the subject identifications so that we could no longer determine which subject belonged in which category. Imagine you are given two sets of 24 labels. The first set has 14 labels with the word "woman" and 10 labels with the word "man." The second set of labels has 11 labels with the word "dead" and 12 labels with the word "alive." Under the null hypothesis, you are allowed to distribute the labels to subjects independently of one another—one label from each of the two sets per subject, please.

There are a total of $\binom{24}{10} = 24!/(10!14!) = 1{,}961{,}256$ ways you could hand out the labels. Here $\binom{14}{10}\binom{10}{1} = 10{,}010$ of the assignments result in tables that are as extreme as our original table (that is, in which 90% of the men survive) and $\binom{14}{11}\binom{10}{0} = 364$ of the assignments result in tables that are more extreme (100% of the men survive). This is a very small fraction of the total, $(10{,}010 + 364)/(1{,}961{,}256) = 0.529\%$, so we conclude that a difference in the survival rates of the two sexes as extreme as the difference we observed in our original table is very unlikely to have occurred by chance alone. We reject the hypothesis that the survival rates for the two sexes are the same and accept the alternative hypothesis that, in this instance at least, men are more likely to profit from treatment.

	Survived	Died	Total
Men	10	0	10
Women	3	11	14
Total	13	11	24

	Survived	Died	Total
Men	8	2	10
Women	5	9	14
Total	13	11	24

In terms of the relative survival rates of the two sexes, the first of these tables is more extreme than our original table. The second is less extreme.

In the preceding example, we tested the hypothesis that survival rates do not depend on sex against the alternative that men diagnosed with cancer are likely to live longer than women similarly diagnosed. We rejected the null hypothesis because only a small fraction of the possible tables were as extreme as the one we observed initially. This is an example of a one-tailed test. But is it the correct test? Is this really the alternative hypothesis we would have proposed if we had not already seen the data? Wouldn't we have been just as likely to reject the null hypothesis that men and women profit equally from treatment if we had observed a table of the following form?

	Survived	Died	Total
Men	0	10	10
Women	13	1	14
Total	13	11	24

Of course we would! In determining the significance level in the present example, we must add together the total number of tables that lie in either of the two extremes, or tails, of the permutation distribution.

The critical values and significance levels are quite different for one-tailed and two-tailed tests and, all too often, the wrong test has been employed in published work. McKinney et al. [1989] reviewed more than 70 articles that appeared in six medical journals. In over half of these articles, Fisher's exact test was applied improperly. Either a one-tailed test had been used when a two-tailed test was called for or the authors of the article simply hadn't bothered to state which test they had used.

Of course, unless you are submitting the results of your analysis to a regulatory agency, no one will know whether you originally intended a one-tailed test or a two-tailed test and subsequently changed your mind. No one will know whether your hypothesis was conceived before you started or only after you examined the data. All you have to do is lie. Just recognize that if you test an after-the-fact hypothesis without identifying it as such, you are guilty of scientific fraud.

When you design an experiment, decide at the same time whether you wish to test your hypothesis against a two-sided or a one-sided alternative. A two-sided alternative dictates a two-tailed test; a one-sided alternative dictates a one-tailed test.

As an example, suppose we decide to do a follow-on study of the cancer registry to confirm our original finding that men diagnosed as having tumors live significantly longer than women similarly diagnosed. In this follow-on study, we have a one-sided alternative. Thus, we would analyze the results using a one-tailed test rather than the two-tailed test we applied in the original study.

Determine beforehand whether your alternative hypotheses are ordered or unordered.

Ordered or Unordered Alternative Hypotheses?

When testing qualities (number of germinating plants, crop weight, etc.) from k samples of plants taken from soils of different composition, it is often routine to use the F-ratio of the analysis of variance. For contingency tables, many routinely use the chi-square test to determine if the differences among samples are significant. But the F-ratio and the chi-square are what are termed omnibus tests, designed to be sensitive to all possible alternatives. As such, they are not particularly sensitive to ordered alternatives such "as more fertilizer, more growth" or "more aspirin, faster relief of headache." Tests for such ordered responses at k distinct treatment levels should properly use the Pitman correlation described by Frank et al. [1978] when the data are measured on a metric scale (e.g., weight of the crop). Tests for ordered responses in $2 \times C$ contingency tables (e.g., number of germinating plants) should use the trend test described by Berger, Permutt, and Ivanova [1998]. We revisit this topic in more detail in the next chapter.

DEDUCTION AND INDUCTION

When we determine a p-value as we did in the example above, we apply a set of algebraic methods and deductive logic to *deduce* the correct value. The deductive process is used to determine the appropriate size of resistor to use in an electric circuit, to determine the date of the next eclipse of the moon, and to establish the identity of the criminal (perhaps from the fact that the dog did not bark on the night of the crime). Find the formula, plug in the values, turn the crank, and out pops the result (or it does for Sherlock Holmes,[4] at least).

[4]See "Silver Blaze" by A. Conan-Doyle, *Strand* magazine, December 1892.

When we assert that for a given population a percentage of samples will have a specific composition, this also is a deduction. But when we make an *inductive* generalization about a population based upon our analysis of a sample, we are on shakier ground. It is one thing to assert that if an observation comes from a normal distribution with mean zero, the probability is one-half that it is positive. It is quite another if, on observing that half of the observations in the sample are positive, we assert that half of all the possible observations that might be drawn from that population will also be positive.

Newton's law of gravitation provided an almost exact fit (apart from measurement error) to observed astronomical data for several centuries; consequently, there was general agreement that Newton's generalization from observation was an accurate description of the real world. Later, as improvements in astronomical measuring instruments extended the range of the observable universe, scientists realized that Newton's law was only a generalization and not a property of the universe at all. Einstein's theory of relativity gives a much closer fit to the data, a fit that has not been contradicted by any observations in the century since its formulation. But this still does not mean that relativity provides us with a complete, correct, and comprehensive view of the universe.

In our research efforts, the only statements we can make with God-like certainty are of the form "Our conclusions fit the data." The true nature of the real world is unknowable. We can speculate, but never conclude.

LOSSES

In our first advanced course in statistics, we read in the first chapter of Lehmann [1986] that the "optimal" statistical procedure would depend on the losses associated with the various possible decisions. But on day one of our venture into the real world of practical applications, we were taught to ignore this principle.

At that time, the only computationally feasible statistical procedures were based on losses that were proportional to the square of the difference between estimated and actual values. No matter that the losses really might be proportional to the absolute value of those differences, or the cube, or the maximum over a certain range. Our options were limited by our ability to compute.

Computer technology has made a series of major advances in the past half century. What 40 years ago required days or weeks to calculate takes only milliseconds today. We can now pay serious attention to this long neglected facet of decision theory: the losses associated with the varying types of decisions.

Suppose we are investigating a new drug. We gather data, perform a statistical analysis, and draw a conclusion. If chance alone is at work yielding exceptional values, and we opt in favor of the new drug, we've made an error. We also make an error if we decide there is no difference and the new drug really is better. These decisions and the effects of making them are summarized in Table 2.1.

We distinguish the two types of error because they have quite different implications as described in Table 2.1. As a second example, Fears, Tarone, and Chu [1977] use

TABLE 2.1. Decision Making Under Uncertainty

The Facts	Our Decision	
	No Difference	Drug Is Better
No Difference	Correct	Type I error: Manufacturer wastes money developing an ineffective drug

Drug Is Better	Type II error:
	Manufacturer misses opportunity
	    for profit
	Public denied access to effective
	    treatment
		Correct

TABLE 2.2. Decision Making Under Uncertainty

The Facts	Fears et al's Decision	
	Not a Carcinogen	Carcinogen

Not a Carcinogen		Type I error:
		Manufacturer misses opportunity
		    for profit
		Public denied access to effective
		    treatment

Carcinogen	Type II error: Patients die; families suffer Manufacturer sued	

permutation methods to assess several standard screens for carcinogenicity. As shown in Table 2.2, their Type I error, a false positive, consists of labeling a relatively innocuous compound as carcinogenic. Such an action means economic loss for the manufacturer and the denial to the public of the compound's benefits. Neither consequence is desirable. But a false negative, a Type II error, is much worse, as it means exposing a large number of people to a potentially lethal compound.

What losses are associated with the decisions you will have to make? Specify them before you begin.

DECISIONS

The primary hypothesis/alternative hypothesis duality is inadequate in most real-life situations. Consider the pressing problems of global warming and depletion of the ozone layer. We could collect and analyze yet another set of data and then, just as

TABLE 2.3. Effect on Global Warming

	President's Decision on Emissions		
The Facts	Reduce Emissions	Gather More Data	Change Unnecessary
Emissions responsible	Global warming slows	Decline in quality of life (irreversible?)	Decline in quality of life
Emissions have no effect	Economy disrupted	Sampling costs	

is done today, make one of three possible decisions: reduce emissions, leave emission standards alone, or sit and wait for more data to come in. Each decision has consequences, as shown in Table 2.3.

As noted at the beginning of this chapter, it's essential we specify in advance the actions to be taken for each potential result. Always suspect are after-the-fact rationales that enable us to persist in a pattern of conduct despite evidence to the contrary. If no possible outcome of a study will be sufficient to change our minds, then we should not undertake such a study in the first place.

Every research study involves multiple issues. Not only might we want to know whether a measurable, biologically (or medically, physically, or sociologically) significant effect takes place, but also the size of the effect and the extent to which it varies from instance to instance. We would also want to know what factors, if any, will modify the size of the effect or its duration.

We may not be able to address all these issues with a single data set. A preliminary experiment might tell us something about the possible existence of an effect, along with rough estimates of its size and variability. Hopefully, we glean enough information to come up with doses, environmental conditions, and sample sizes to apply in collecting and evaluating the next data set. A list of possible decisions after the initial experiment includes "abandon this line of research," "modify the environment and gather more data," and "perform a large, tightly controlled, expensive set of trials." Associated with each decision is a set of potential gains and losses. Common sense dictates that we construct a table similar to Table 2.1 or 2.2 before we launch a study.

For example, in clinical trials of a drug, we might begin with some animal experiments, then progress to Phase I clinical trials in which, with the emphasis on safety, we look for the maximum tolerable dose. Phase I trials generally involve only a small number of subjects and a one-time or short-term intervention. An extended period of several months may be used for follow-up purposes. If no adverse effects are observed, we might decide to pursue a Phase II set of trials in the clinic, in which our objective is to determine the minimum effective dose. Obviously, if the minimum effective dose is greater than the maximum tolerable dose, or if some dangerous side effects are observed that we didn't observe in the first set of trials, we'll abandon the drug and go on to some other research project. But if the signs are favorable, then and only then will we go to a set of Phase III trials involving a large number of subjects observed over an extended time period. Then and only then will we hope to get the answers to all our research questions.

> Before you begin, list all the consequences of a study and all the actions you might take. Persist only if you can add to existing knowledge.

TO LEARN MORE

For more thorough accounts of decision theory, the interested reader is directed to Berger [1986], Blyth [1970], Cox [1958], DeGroot [1970], and Lehmann [1986]. For an applied perspective, see Clemen [1991], Berry [1995], and Sox, Blatt, Higgins, and Marton [1988].

Over 300 references warning of the misuse of null hypothesis testing can be accessed online at http://www.cnr.colostate.edu/~ anderson/thompson1.html. Alas, the majority of these warnings are ill informed, stressing errors that will not arise if you proceed as we recommend and place the emphasis on the why, not the what, of statistical procedures. Use statistics as a guide to decision making rather than a mandate.

Neyman and Pearson [1933] first formulated the problem of hypothesis testing in terms of two types of error. Extensions and analyses of their approach are given by Lehmann [1986] and Mayo [1996]. Their approach has not gone unchallenged as seen in Berger and Berry [1988], Berger and Selike [1987], Berkson [1942], Morrison and Henkel [1970], Savage [1972], Schmidt [1996], Seidenfeld [1979], and Sterne, Smith, and Cox [2001]. Hunter and Schmidt [1997] list and dismiss many of their objections.

Guthery, Lusk, and Peterson [2001] and Rozeboom [1960] are among those who have written about the inadequacy of the null hypothesis.

For more guidelines on formulating meaningful primary hypotheses, see Selike, Bayarri, and Berger [2001]. Clarity in hypothesis formulation is essential; ambiguity can only yield controversy; see, for example, Kaplan [2001].

Venn [1866] and Reichenbach [1949] are among those who've attempted to construct a mathematical bridge between what we observe and the reality that underlies our observations. Such efforts to the contrary, extrapolation from the sample to the population is not merely a matter of applying Sherlock Holmes-like deductive logic but entails a leap of faith. A careful reading of Locke [1700], Berkeley [1710], Hume [1748], and Lonergan [1992] is an essential prerequisite to the application of statistics.

See also Buchanan-Wollaston [1935], Cohen [1990], Copas [1997], Lindley [2000].

3

COLLECTING DATA

GIGO: Garbage in, garbage out.

Fancy statistical methods will not rescue garbage data.
                                    —Course notes of Raymond J. Carroll [2001]

The vast majority of errors in statistics (and, not incidentally, in most human endeavors) arise from a reluctance (or even an inability) to plan. Some demon (or demonic manager) seems to be urging us to cross the street before we've had the opportunity to look both ways. Even on those rare occasions when we do design an experiment, we seem more obsessed with the mechanics than with the underlying concepts.

In this chapter we review the fundamental concepts of experimental design, the choice of primary and secondary variables, the selection of mesurement devices, the determination of sample size, and the assumptions underlie most statistical procedures along with the precautions necessary to ensure they are satisfied and that the data you collect will be representative of the population as a whole. We do not intend to replace a textbook on experiment or survey design but rather to supplement it, providing examples and solutions that are often neglected in courses on the subject.

PREPARATION

The first step in data collection is to have a clear, preferably written statement of your objectives. In accordance with Chapter 1, you will have defined the population or populations from which you intend to sample and identified the characteristics of these populations that you wish to investigate.

Common Errors in Statistics (and How to Avoid Them), Third Edition. Edited by P. I. Good and J. W. Hardin
Copyright © 2009 John Wiley & Sons, Inc.

You developed one or more well-formulated hypotheses (the topic of Chapter 2) and have some idea of the risks you will incur should your analysis of the collected data prove to be erroneous. You will need to decide what you wish to observe and measure and how you will go about observing it. We refer here not only to the primary variables or endpoints, but also to the secondary variables or cofactors that may influence the former's values.

A good practice is to draft the analysis section of your final report based on the conclusions you would like to make. What information do you need to justify these conclusions? All such information must be collected.

The next section is devoted to the choice of response variables and measuring devices, followed by sections on determining sample size and preventive steps to ensure that your samples will be analyzable by statistical methods.

RESPONSE VARIABLES

Know what you want to measure. If you don't collect the values of cofactors, you will be unable to account for them later.

As whiplash injuries are a common consequence of rear-end collisions, there is an extensive literature on the subject. Any physician will tell you that the extent and duration of such injuries depend upon the sex, age, and physical condition of the injured individual as well as any prior injuries the individual may have suffered. Yet, we found article after article that failed to account for these factors; for example, Krafft et al. [2002], Kumar, Ferrari, and Narayan [2005], and Tencer, Sohail, and Kevin [2001] did not report the sex, age, or prior injuries of their test subjects.

Will you measure an endpoint such as death or a surrogate such as the presence of human immunodeficiency virus (HIV) antibodies? A good response variable takes values over a sufficiently large range that they discriminate well [Bishop and Talbot, 2001].

The regression slope describing the change in systolic blood pressure (in mm Hg) per 100 mg of calcium intake is strongly influenced by the approach used for assessing the amount of calcium consumed [Cappuccio et al., 1995]. The association is small and only marginally significant with diet histories [slope -0.01 (-0.003 to -0.016)] but large and highly significant when food frequency questionnaires are used [-0.15 (-0.11 to -0.19)]. With studies using 24-hour recall, an intermediate result emerges [-0.06 (-0.09 to -0.03)]. Diet histories assess patterns of usual intake over long periods of time and require an extensive interview with a nutritionist, whereas 24-hour recall and food frequency questionnaires are simpler methods that reflect current consumption [Block, 1982].

Before we initiate data collection, we must have a firm idea of what we will measure and how we will measure it. A good response variable

- Is easy to record—imagine weighing a live pig.
- Can be measured objectively on a generally accepted scale.

- Is measured in appropriate units.
- Takes values over a sufficiently large range that discriminates well.
- Is well-defined. A patient is not "cured" but may be "discharged from the hospital" or "symptom-free for a predefined period."
- Has constant variance over the range used in the experiment [Bishop and Talbot, 2001].

Collect exact values whenever possible.

A second fundamental principle is also applicable to both experiments and surveys: Collect exact values whenever possible. Worry about grouping them in intervals or discrete categories later.

A long-term study of buying patterns in New South Wales illustrates some of the problems caused by grouping prematurely. At the beginning of the study, the decision was made to group the incomes of survey subjects into categories: under $20,000, $20,000 to $30,000, and so forth. Six years of steady inflation later, the organizers of the study realized that all the categories had to be adjusted. An income of $21,000 at the start of the study would purchase only $18,000 worth of goods and housing at the end (see Fig. 3.1). The problem was that those persons surveyed toward the end had filled out forms with exactly the same income categories. Had income been tabulated to the nearest dollar, it would have been easy to correct for increases in the cost of living and convert all responses to the same scale. But the

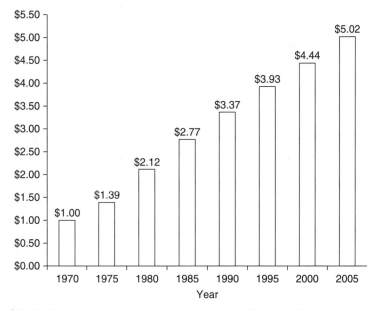

Figure 3.1. Equivalent purchasing powers over time using Consumer Price Index calculations. Each year shows the cost of the equivalent goods/services.

study designers hadn't considered these issues. A precise and costly survey was now a matter of guesswork.

You can always group your results (and modify your groupings) after a study is completed. If after-the-fact grouping is a possibility, your design should state how the grouping will be determined; otherwise, there will be the suspicion that you chose the grouping to obtain desired results.

Experiments

Measuring devices differ widely both in what they measure and in the precision with which they measure it. As noted in the next section of this chapter, the greater the precision with which measurements are made, the smaller the sample size required to reduce both Type I and Type II errors below specific levels.

All measuring devices have both linear and nonlinear ranges; the sensitivity, accuracy, and precision of the device are all suspect for both very small and very large values. Your measuring device ought to be linear over the entire range of interest.

Before you rush out and purchase the most expensive and precise measuring instruments on the market, consider that the total cost C of an experimental procedure is $S + nc$, where n is the sample size and c is the cost per unit sampled.

The startup cost S includes the cost of the measuring device; c consists of the cost of supplies and personnel. The latter includes not only the time spent on individual measurements but also in preparing and calibrating the instrument for use.

Less obvious factors in the selection of a measuring instrument include the impact on the subject, reliability (personnel costs continue even when an instrument is down), and reusability in future trials. For example, one of the advantages of the latest technology for blood analysis is that less blood needs to be drawn from patients. Less blood means happier subjects, mean fewer withdrawals, and a smaller initial sample size.

Surveys

While no scientist would dream of performing an experiment without first mastering all the techniques involved, an amazing number will blunder into the execution of large-scale and costly surveys without a preliminary study of all the collateral issues a survey entails.

We know of one institute that mailed out some 20,000 questionnaires (didn't the postal service just raise its rates again?) before discovering that half of the addresses were in error and that the vast majority of the remainder were being discarded unopened before prospective participants had even read the "sales pitch."

Fortunately, there are textbooks, such as those of Bly [1990, 1996] that will tell you how to word a sales pitch and the optimal colors and graphics to use along with the wording. They will tell you what "hooks" to use on the envelope to ensure attention to the contents and what premiums to offer to increase participation.

There are other textbooks, such as those of Converse and Presser [1986], Fowler and Fowler [1995], and Schroeder [1987], to assist you in wording questionnaires and in pretesting questions for ambiguity before you begin. We have only three cautions to offer:

1. Be sure your questions don't reveal the purpose of your study; otherwise, respondents will shape their answers to what they perceive to be your needs. Contrast "How do you feel about compulsory pregnancy?" with "How do you feel about abortions?"
2. With populations that are ever more heterogeneous, questions that work with some ethnic groups may repel others [see, e.g., Choi, 2000].
3. Be sure to include a verification question or two. For example, in March 2000, the U.S. Census Current Population Survey added an experimental health insurance verification question. Anyone who did not report any type of health insurance coverage was asked an additional question about whether or not he or she was, in fact, uninsured. Those who reported that they were insured were then asked what type of insurance covered them.

Recommended are Web-based surveys with initial solicitation by mail (letter or post card) and email. Not only are both costs and time to completion cut dramatically, but the proportion of missing data and incomplete forms is substantially reduced. Moreover, Web-based surveys are easier to monitor and forms may be modified on the fly. Web-based entry also offers the possibility of displaying the individual's prior responses during follow-up surveys.

Three other precautions can help ensure the success of your survey:

1. Award premiums only for fully completed forms.
2. Continuously tabulate and monitor submissions; don't wait to be surprised.
3. A quarterly newsletter sent to participants will substantially increase retention (and help you keep track of address changes).

BEWARE OF HOLES IN THE INSTRUCTIONS

The instructions for Bumbling Pharmaceutical's latest set of trials seemed almost letter perfect. At least they were lengthy and complicated enough that they intimidated anyone who took the time to read them. Consider the following, for example:

"All patients will have follow-up angiography at eight $\pm$ 0.5 months after their index procedure. Any symptomatic patient will have follow-up angiograms any time it is clinically indicated. In the event that repeat angiography demonstrates restenosis in association with objective evidence of recurrent ischemia between zero and six months, that angiogram will be analyzed as the follow-up angiogram. An angiogram performed for any reason that doesn't show restenosis will qualify as a follow-up angiogram only if it is performed at least four months after the index intervention.

In some cases, recurrent ischemia may develop within 14 days after the procedure. If angiography demonstrates a significant residual stenosis ($>$50%) and if further intervention is performed, the patient will still be included in the follow-up analyses that measure restenosis."

Now, that's comprehensive. Isn't it? Just a couple of questions: If a patient doesn't show up for their [sic] 8-month follow-up exam but does appear at six months and one year, which angiogram should be used for the official reading? If a patient develops recurrent ischemia 14 days after the procedure and a further intervention is performed, do we reset the clock to zero days?

Alas, these holes in the protocol were discovered by Bumbling's staff only *after* the data were in hand and they were midway through the final statistical analysis. Have someone who thinks like a programmer (or, better still, have a computer) review the protocol before it is finalized.

Source: Reprinted with permission from *A Manager's Guide to the Design and Conduct of Clinical Trials*, 2nd ed, Hoboken, NJ: Wiley, 2005.

DETERMINING SAMPLE SIZE

Determining the optimal sample size is simplicity itself once we specify all of the following:

- Smallest effect of clinical or experimental significance.
- Desired power and significance level.
- Distributions of the observables.
- Statistical test(s) that will be employed.
- Whether we will be using a one-tailed or a two-tailed test.
- Anticipated losses due to nonresponders, noncompliant participants, and dropouts.

What could be easier?

Power and Significance Level

Sample size must be determined for each experiment; there is no universally correct value. We need to understand and use the relationships among effect size, sample size, significance level, power, and the precision of our measuring instruments.

If we increase the precision (and hold all other parameters fixed), we can decrease the required number of observations. Decreases in any or all of the intrinsic and extrinsic sources of variation will also result in a decrease in the required number.

The smallest effect size of practical interest should be determined first through consultation with one or more domain experts. The smaller this value, the greater the number of observations that will be required.

TABLE 3.1. Ingredients in a Sample Size Calculation

Smallest Effect Size of Practical Interest	Definition
Type I error (α)	Probability of falsely rejecting the hypothesis when it is true.
Type II error ($1 - \beta[A]$)	Probability of falsely accepting the hypothesis when an alternative hypothesis A is true. This depends on alternative A.
Power $= \beta[A]$	Probability of correctly rejecting the hypothesis when an alternative hypothesis A is true. This depends on alternative A.
Distribution functions	$F[(x - \mu)\sigma]$ (e.g., normal distribution).
Location parameters	For both the hypothesis and the alternative hypothesis, μ_1, μ_2.
Scale parameters	For both the hypothesis and the alternative hypothesis, σ_1, σ_2.
Sample sizes	May be different for different groups in an experiment with more than one group.

If we permit a greater number of Type I or Type II errors (and hold all other parameters fixed), we can decrease the required number of observations.

Explicit formulas for power and sample size are available when the underlying observations are binomial, the results of a counting or Poisson process, time-to-event data, normally distributed, or ordinal with a limited number of discrete values (fewer than seven), and/or the expected proportion of cases at the boundaries is high (scoring 0 or 100). For the first four, several off-the-shelf computer programs including nQuery Advisor™, Pass 2005™, Power and Precision™, and StatXact™ are available to do the calculations for us. For the ordinal data, use the method of Whitehead [1993].

During a year off from Berkeley's graduate program to work as a statistical consultant, with a course from Erich Lehmann in testing hypotheses fresh under his belt, Phillip Good would begin by asking all clients for their values of α and β. When he received only blank looks in reply, he would ask them about the relative losses they assigned to Type I and Type II errors, but this only added to their confusion. Here are some guidelines; just remember this is all they are: guidelines, not universal truths. Strictly speaking, the significance level and power should be chosen so as to minimize the overall cost of any project, balancing the cost of sampling with the costs expected from Type I and Type II errors.

The environment in which you work should determine your significance level and power.

A manufacturer preparing to launch a new product line or a pharmaceutical company conducting research for promising compounds typically adopt a three-way decision procedure: If the observed p-value is less than 1%, they go forward with the project. If the p-value is greater than 20%, they abandon it. And if the p-value lies in the gray area in between, they arrange for additional surveys or experiments.

A regulatory commission like the Food and Drug Administration (FDA) that is charged with oversight responsibility must work at a fixed significance level, typically

5%. In the case of unwanted side effects, the FDA may also require a certain minimum power, usually 80% or better. The choice of a fixed significance level ensures consistency in both result and interpretation as the agency reviews the findings from literally thousands of tests.

If forced to pull numbers out of a hat, we'd choose $\alpha = 20\%$ for an initial trial and a sample size of 6 to 10. If we had some prior information in hand, we'd choose $\alpha = 5\%$ to 10% and $\beta = 80\%$ to 90%.

When using one of the commercially available programs to determine sample size, we also need to have some idea of the population proportion (for discrete counts) or the location (mean) and scale parameter (variance) (for continuous measurements) both when the primary hypothesis is true and when an alternative hypothesis is true. Since there may well be an infinite number of alternatives in which we are interested, power calculations should be based on the worst case or boundary value. For example, if we are testing a binomial hypothesis $p = \frac{1}{2}$ against the alternatives $p \leq \frac{2}{3}$, we would assume that $p = \frac{2}{3}$.

A recommended rule of thumb is to specify as the alternative the smallest effect that is of practical significance.

When determining sample size for data drawn from the binomial or any other discrete distribution, one should always display the power curve. The explanation lies in the sawtoothed nature of the curve [Chernick and Liu, 2002]; see Fig. 3.2. As a result of inspecting the power curve visually, you may come up with a less expensive solution than your software.

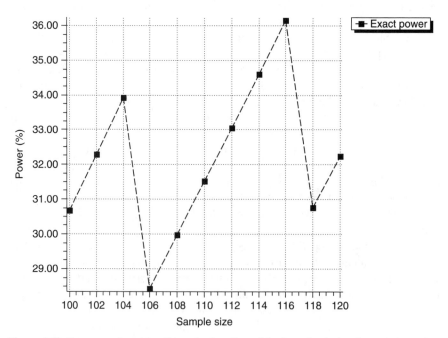

Figure 3.2. Power as a function of sample size. Test of the hypothesis that the prevalence in a population of a characteristic is 0.46 rather than 0.40.

**TABLE 3.2. Bootstrap Estimates of Type I and
Type II Error**

	Test Mean < 6.0	
Sample Size	α	Power
3	0.23	0.84
4	0.04	0.80
5	0.06	0.89

If the data do not come from a well-tabulated distribution, then one might use a bootstrap to estimate the power and significance level.

In preliminary trials of a new device, test results of 7.0 were observed in 11 out of 12 cases and 3.3 in 1 out of 12 cases. Industry guidelines specified that any population with a mean test result greater than 5.0 would be acceptable. A worst case or boundary-value scenario would include one in which the test result was 7.0 $\frac{3}{7}$th of the time, 3.3 $\frac{3}{7}$th of the time, and 4.1 $\frac{1}{7}$th of the time.

The statistical procedure required us to reject if the sample mean of the test results were less than 6.0. To determine the probability of this event for various sample sizes, we took repeated samples with replacement from the two sets of test results. Some bootstrap samples consisted of all 7s; others, taken from the worst-case distribution, consisted only of 3.3s. Most were a mixture. Table 3.2 illustrates the results; for example, in our trials, 23% of the bootstrap samples of size 3 from our starting sample of test results had medians less than 6. If we drew our bootstrap samples from the hypothetical "worst-case" population, then 84% had medians less than 6.

If you want to try your hand at duplicating these results, simply take the test values in the proportions observed, stick them in a hat, draw out bootstrap samples with replacement several hundred times, compute the sample means and record the results. Or you could use the R or Stata bootstrap procedure as we did.[1]

DON'T LET THE SOFTWARE DO YOUR THINKING FOR YOU

Many researchers today rely on menu-driven software to do their power and sample size calculations. Most such software comes with default settings—for example, alpha = 0.05, tails = 2, settings that are readily altered, if, that is, investigators bothers to take the time.

Among the errors made by participants in a recent workshop on sample size determination was letting the software select a two-sample, two-tailed test for the hypothesis that 50% or less of subjects would behave in a certain way versus the alternative that 60% or more of them would.

---

[1]Chapters 5–8 provide more information on the bootstrap and its limitations.

SEQUENTIAL SAMPLING

Determing sample size as we go (sequential sampling), rather than making use of a pre-determined sample size, can have two major advantages:

1. Fewer samples
2. Earlier decisions

When our experiments are destructive in nature (as in testing condoms) or may have an adverse effect upon the experimental subject (as in clinical trials) we would prefer not to delay our decisions untill some fixed sample size has been reached.

Figure 3.3 depicts a sequential trial of a new vaccine after eight patients who had received either the vaccine or an innocuous saline solution developed the disease. Each time a control patient came down with the disease, the jagged line was extended to the right. Each time a patient who had received the experimental vaccine came down with the disease, the jagged line was extended upward one notch. This experiment will continue until either of the following occurs:

A. the jagged line crosses the lower boundary—in which case, we will stop the experiment, reject the null hypothesis and immediately put the vaccine into production,
B. the jagged line crosses the upper boundary—in which case, we will stop the experiment, accept the null hypothesis and abandon further work with this vaccine.

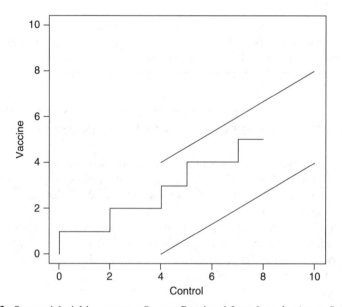

Figure 3.3. Sequential trial in progress. *Source*: Reprinted from *Introduction to Statistics Via Resampling Methods and R/SPlus* with the permission of John Wiley & Sons, Inc.

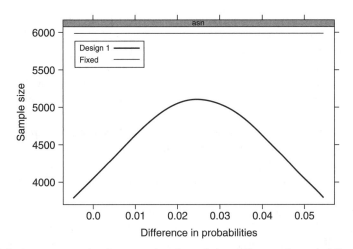

Figure 3.4. Average sample size as a function of the difference in probability. *Source*: Reprinted from *Introduction to Statistics Via Resampling Methods and R/SPlus* with the permission of John Wiley & Sons, Inc.

What Abraham Wald [1950] showed in his pioneering research was that on the average the resulting sequential experiment *would require many fewer observations* whether or not the vaccine was effective than would a comparable experiment of fixed sample size.

If the treatment is detrimental to the patient, we are likely to hit one of the lower boundaries early. If the treatment is far more efficacious than the control, we are likely to hit an upper boundary early. Even if the true difference is right in the middle between our two hypotheses, for example, because the treatment is only 2.5% better when the alternative hypothesis is that it is 5% better, we may stop early on occasion. Figure 3.4 shows the average sample size as a function of the difference in the probabilities of success for each treatment. When this difference is less than 0% or greater than 5%, we'll need about 4000 observations on average before stopping. Even when the true difference is right in the middle, we will stop after about 5000 observations, on average. In contrast, a fixed-sample design requires nearly 6000 observations for the same Type I error and power.

Warning: simply performing a standard statistical test after each new observation as if the sample size were fixed will lead to inflated values of Type I error . The boundaries depicted in Figure 3.3 were obtained using formulas specific to sequential design. Not surprisingly, these formulas require us to know every one of the same factors we needed to determine the number of samples when the experiment is of fixed size.

ONE-TAIL OR TWO?

A one-sided alternative ("Regular brushing with a flouride toothpaste will reduce cavities") requires a one-tailed or one-sided test. A two-sided alternative ("Which brand ought one buy?") requires a two-tailed test.

But it often can be difficult in practical situations to decide whether the real alternatives are one-sided or two-sided. Moyé [2000] provides a particularly horrifying illustration in his textbook *Statistical Reasoning in Medicine* (pp. 145–148) which, in turn, was extracted from Thomas Moore's *Deadly Medicine* [1995, pp. 203–204]. It concerns a study of cardiac arrhythmia suppression, in which a widely used but untested therapy was at last tested in a series of controlled (randomized, double-blind) clinical trials [Greene et al., 1992].

The study had been set up as a sequential trial. At various checkpoints, the trial would be stopped if efficacy was demonstrated or if it became evident that the treatment was valueless. Though no preliminary studies had been made, the investigators did not plan for the possibility that the already widely used but untested treatment might be harmful. It was.

Fortunately, clinical trials have multiple endpoints, an invaluable resource, if the investigators choose to look at the data. In this case, a Data and Safety Monitoring Board (consisting of scientists not directly affiliated with the trial or the trial's sponsors) noted that of 730 patients randomized to the active therapy, 56 died, while of 725 patients randomized to placebo, 22 died. They felt free to perform a two-sided test despite the original formulation of the problem as one-sided.

Prepare for Missing Data

The relative ease with which a program like Stata or Power and Precision can produce a sample size may blind us to the fact that the number of subjects with which we begin a study may bear little or no relation to the number with which we conclude it.

A midsummer hailstorm, an early frost, or an insect infestation can lay waste to all or part of an agricultural experiment. In the National Institute of Aging's first years of existence, a virus wiped out a primate colony, destroying a multitude of experiments in progress.

Large-scale clinical trials and surveys have a further burden: the subjects themselves. Potential subjects can and do refuse to participate. (Don't forget to budget for a follow-up study, bound to be expensive, of responders versus nonresponders.) Worse, they agree to participate initially, then drop out at the last minute (see Fig. 3.5).

They move without leaving a forwarding address before a scheduled follow-up, or they simply don't bother to show up for an appointment. Thirty percent of the patients who had received a lifesaving cardiac procedure failed to follow up with their physician. (We can't imagine not going to see our surgeon after such a procedure, but then we guess we're not typical.)

The key to a successful research program is to plan for such dropouts in advance and to start the trials with some multiple of the number required to achieve a given power and significance level.

In a recent attempt to reduce epidemics at its training centers, the U.S. Navy vaccinated 50,000 recruits with an experimental vaccine and 50,000 others with a harmless saline solution. But at the halfway mark, with 50,000 inoculated and 50,000 to go, fewer than 500 had contracted the disease. The bottom line: It's the sample you end with, not the sample you begin with, that determines the power of your tests.

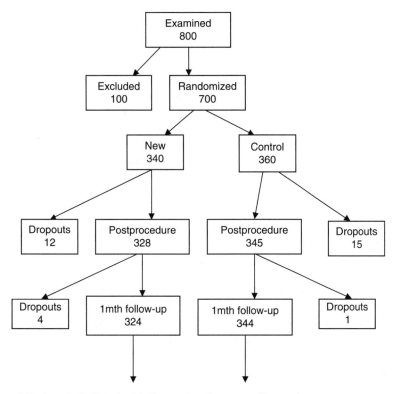

Figure 3.5. A typical clinical trial. Dropouts and noncompliant patients occur at every stage. *Source*: Reprinted from the *Manager's Guide to Design and Conduct of Clinical Trials* with the permission of John Wiley & Sons, Inc.

Nonresponders

An analysis of those who did not respond to a survey or a treatment can sometimes be as informative or more informative than the survey itself. See, for example, the work of Mangel and Samaniego [1984] as well as the sections on the Behrens–Fisher problem and on the premature drawing of conclusions in Chapter 5. Be sure to incorporate in your sample design and in your budget provisions for sampling nonresponders.

Sample from the Right Population

Be sure that you are sampling from the population as a whole rather than from an unrepresentative subset of the population. The most famous blunder along these lines was basing the forecast of Dewey defeating Truman in the 1948 U.S. presidential election on a telephone survey; those who owned a telephone and responded to the survey favored Dewey; those who voted did not.

An economic study may be flawed because we have overlooked the homeless. This was among the principal arguments that New York City and Los Angeles advanced against the federal government's use of the 1990 and 2000 censuses to determine the basis for awarding monies to cities.[2]

An astrophysical study was flawed because it overlooked galaxies whose central surface brightness was very low.[3] The FDA's former policy of permitting clinical trials to be limited to men was just plain foolish.

In contributing to a plaintiff's lawsuit following a rear-end collision, Dr. Good noted that while the plaintiff was in her 50s and had been injured previously, the studies relied on by the defendant's biomechanical expert involved only much younger individuals with no prior history of injury.

Plaguing many surveys are the uncooperative and the non-responders. Invariably, follow-up surveys of these groups show substantial differences from those who responded readily the first time. These follow-up surveys aren't inexpensive—compare the cost of mailing out a survey to that of telephoning or making face-to-face contact with a nonresponder. But if one doesn't follow up, one may get a completely unrealistic picture of how the population as a whole would respond.[4]

FUNDAMENTAL ASSUMPTIONS

Most statistical procedures rely on two fundamental assumptions: that the observations are independent of one another and that they are identically distributed. If your methods of collection fail to honor these assumptions, then your analysis must also fail.

Independent Observations

To ensure the independence of responses in a return-by-mail or return-by-Web survey, no more than one form per household should be accepted. If a comparison of the responses within a household is desired, then the members of the household should be interviewed separately, outside of each other's hearing, and with no opportunity to discuss the survey in between. People care what other people think, and when asked about an emotionally charged topic, they may or may not tell the truth. In fact, they are unlikely to tell the truth if they feel that others may overhear or somehow learn of their responses.

To ensure independence of the observations in an experiment, determine in advance what constitutes the *experimental unit*.

In the majority of cases, the unit is obvious: one planet means one position in space, one container of gas means one volume and pressure to be recorded, one runner on one fixed race course means one elapsed time.

In a clinical trial, each patient corresponds to a single set of observations—or does she? Suppose we are testing the effects of a topical ointment on pink eye. Is each eye a separate experimental unit or is each patient?

[2]*City of New York v. Dept. of Commerce*, 822 F. Supp. 906 (E.D.N.Y., 1993).
[3]Bothun [1998, p. 249].
[4]

It is common in toxicology to examine a large number of slides. But regardless of how many are examined in the search for mutagenic and toxic effects, if all slides come from a single treated animal, then the total size of the sample is one.

We may be concerned with the possible effects a new drug might have on a pregnant woman and, as critically, on her children. In our preliminary tests, we'll be working with mice. Is each fetus in the litter a separate experimental unit or is each mother?

If the mother is the one treated with the drug, then the mother is the experimental unit, not the fetus. A litter of six or seven corresponds only to a sample of size one.

As for the topical ointment, while more precise results might be obtained by treating only one eye with the new ointment and recording the subsequent difference in appearance between the treated and untreated eyes, each patient still yields only one observation, not two.

Identically Distributed Observations

If you change measuring instruments during a study or change observers, then you will have introduced an additional source of variation and the resulting observations will not be identically distributed.

The same problems will arise if you discover during the course of a study (as is often the case) that a precise measuring instrument is no longer calibrated and readings have drifted. To forestall this, any measuring instrument should have been exposed to an extensive burn-in before the start of a set of experiments and should be recalibrated as frequently as the results of the burn-in or prestudy period dictate.

Similarly, one doesn't mail out several thousand copies of a survey before performing an initial pilot study to weed out or correct ambiguous and misleading questions.

The following groups are unlikely to yield identically distributed observations: the first to respond to a survey, those who respond only after being offered an inducement, and nonresponders.

EXPERIMENTAL DESIGN

Statisticians have found three ways of coping with individual-to-individual and observer-to-observer variation:

1. *Controlling.* The fewer the extrinsic sources of variation, the smaller the sample size required. Make the environment for the study—the subjects, the manner in which the treatment is administered, the manner in which the observations are obtained, the apparatus used to make the measurements, and the criteria for interpretation—as uniform and homogeneous as possible.

2. *Blocking.* A clinician might stratify the population into subgroups based on such factors as age, sex, race, and the severity of the condition, restricting comparisons to individuals who belong to the same subgroup. An agronomist would want to stratify on the basis of soil composition and environment. (Blocking can also be performed *after* the experiment for the purpose of analysis but *only* if you have taken the time to record the blocking variable.)

3. *Randomizing.* This involves randomly assigning patients to treatment within each subgroup so that the innumerable factors that can neither be controlled nor observed directly are as likely to influence the outcome of one treatment as another.

Steps 1 and 2 are trickier than they appear at first glance. Do the phenomena under investigation depend upon the time of day, as with body temperature and the incidence of mitosis, or upon the day of the week, as with retail sales and the daily mail? Will the observations be affected by the sex of the observer? Primates (including you) and hunters (tigers, mountain lions, domestic cats, dogs, wolves, and so on) can readily detect the observer's sex.[5]

Blocking may be mandatory, as even a randomly selected sample may not be representative of the population as a whole. For example, if a minority comprises less than 10% of a population, then a jury of 12 persons selected at random from that population will fail to contain a single member of that minority at least 28% of the time.

Groups to be compared may differ in other important ways even before any intervention is applied. These baseline imbalances cannot be attributed to the interventions, but they can interfere with and overwhelm the comparison of the interventions.

One good after-the-fact solution is to break the sample itself into strata (men, women, Hispanics) and to extrapolate separately from each stratum to the corresponding subpopulation from which the stratum is drawn.

The size of the sample we take from each block or stratum need not, and in some instances should not, reflect the block's proportion in the population. The latter exception arises when we wish to obtain separate estimates for each subpopulation. For example, suppose we are studying the health of Marine recruits and we wish to obtain separate estimates for male and female Marines as well as for Marines as a group. If we want to establish the incidence of a relatively rare disease, we will need to oversample female recruits to ensure that we obtain a sufficiently large number. To obtain a rate R for *all* Marines, we would then take the weighted average $p_F R_F + p_M R_M$ of the separate rates for each gender, where the proportions p_M and p_F are those of males and females in the *entire* population of Marine recruits.

Are the Study Groups Comparable?

Fujita et al. [2000] compared the short-term effect of (AAACa) and ($CaCO_3$) on bone density in humans. But at the start of the experiment, the bone densities of the $CaCO_3$ group were significantly greater than those of the AAACa group, and the subjects were significantly younger. Thus, the reported changes in bone density could as easily be attributed to differences in age and initial bone density as to differences in the source of supplemental calcium. Clearly, the subjects ought to have been blocked by age and initial bone density before they were randomized to treatment.

---

[5]The hair follicles of redheads, natural rather than dyed, are known to secrete a prostaglandin similar to an insect pheromone.

FOUR GUIDELINES

In the next few sections on experimental design, we may well be preaching to the choir, for which we apologize. But there is no principle of experimental design, however obvious, however intuitive, that someone will not argue can be ignored in his or her special situation:

- Physicians feel that they should be allowed to select the treatment that will best affect their patient's condition (but who is to know in advance what this treatment is?).
- Scientists eject us from their laboratories when we suggest that only the animal caretakers should be permitted to know which cage houses the control animals.
- Engineers at a firm that specializes in refurbishing medical devices objected when Dr. Good suggested that they purchase and test some new equipment for use as controls. "But that would cost a fortune."

The statistician's lot is not a happy one. The opposite sex ignores us because we are boring,[6] and managers hate us because all of our suggestions seem to require an increase in the budget. But controls will save money in the end. Blinding is essential if our results are to have credence, and care in treatment allocation is mandatory if we are to avoid bias.

Randomize

Permitting treatment allocation by either experimenter or subject will introduce bias. On the other hand, if a comparison of baseline values indicates too wide a difference between the various groups in terms of concomitant variables, then you will either need to rerandomize or to stratify the resulting analysis. Be proactive: Stratify before you randomize, randomizing separately within each stratum.

The efforts of Fujita et al. [2000] were doomed before they started, as the placebo-treated group was significantly younger (6 subjects of 50 ± 5 years of age) than the group that had received the treatment of greatest interest (10 subjects of 60 ± 4 years of age).

On the other hand, the study employing case controls conducted by Roberts et al. [2007] could have been rescued had they simply included infant sex as one of the matching variables. For while 85% of the cases of interest were male, only 51% of the so-called matched case controls were of that sex.

Use Controls

To guard against the unexpected, as many or more patients should be assigned to the control regimen as are assigned to the experimental one. This sounds expensive, and it is. But problems happen. You get the flu. You get a headache or the runs. You have a

---

[6]Dr. Good told his wife that he was an author; it was the only way he could lure someone that attractive to his side. Dr. Hardin is still searching for an explanation for his own good fortune.

series of colds that blend into each other until you can't remember the last time you were well. So, you blame your silicone implants. Or, if you are part of a clinical trial, you stop taking the drug. It's in these and similar instances that experimenters are grateful that they've included controls. Because when the data are examined, experimenters learn that as many of the control patients came down with the flu as those who were on the active drug. They also learn that the women without implants had exactly the same incidence of colds and headaches as those with implants.

Reflect on the consequences of not using controls. The first modern silicone implants (Dow Corning's Silastic mammary prosthesis) were placed in 1962. In 1984, a jury awarded $2 million to a recipient who complained of problems resulting from the implants. Award after award followed, the largest being more than $7 million. A set of controlled, randomized trials was finally initiated in 1994. The verdict: Silicon implants have no adverse effects on recipients. Tell this to the stockholders of bankrupt Dow Corning.

Use Positive Controls

There is no point in conducting an experiment if you already know the answer.[7] The use of a positive control is always to be preferred. A new anti-inflammatory should be tested against aspirin or ibuprofen. And there can be no justification whatever for the use of placebo in the treatment of a life-threatening disease [Barbui et al., 2000; Djulbegovic et al., 2000].

Use Blind Observers

Observers should be blinded to the treatment allocation.

Patients often feel better solely because they think they ought to feel better. A drug may not be effective if the patient is aware that it is the old or less favored remedy. Nor is the patient likely to keep taking a drug on schedule if she feels that it contains nothing of value. She is also less likely to report any improvement in her condition if she feels that the doctor has done nothing for her. Conversely, if a patient is informed that he has the new treatment, he may think it necessary to "please the doctor" by reporting some diminishment in symptoms. These behavioral phenomena are precisely the reason that clinical trials must include a control.

A double-blind study in which neither the physician nor the patient knows which treatment is received is preferable to a single-blind study in which only the patient is kept in the dark [Ederer, 1975; Chalmers et al., 1983; Vickers et al., 1997].

Even if a physician has no strong feelings one way or the other concerning a treatment, she may tend to be less conscientious about examining patients she knows belong to the control group. She may have other unconscious feelings that influence her work with the patients. Exactly the same caveats apply in work with animals and plants; units subjected to the existing, less important treatment may be handled more carelessly or examined less thoroughly.

---

[7]The exception being to satisfy a regulatory requirement.

We recommend that you employ two or even three individuals: one to administer the intervention, one to examine the experimental subject, and a third to observe and inspect collateral readings such as angiograms, laboratory findings, and x-rays that might reveal the treatment.

Conceal Treatment Allocation

Without allocation concealment, selection bias can invalidate study results [Schulz, 1995; Berger and Exner, 1999]. If an experimenter could predict the next treatment to be assigned, he might exercise an unconscious bias in the treatment of that patient; he might even defer enrollment of a patient he considers less desirable. In short, randomization alone, without allocation concealment, is insufficient to eliminate selection bias and ensure the internal validity of randomized clinical trials.

Lovell et al. [2000] describe a study in which four patients were randomized to the wrong stratum and, in two cases, the treatment received was reversed. For an excruciatingly (and embarrassingly) detailed analysis of this experiment by an FDA regulator, go to http://www.fda.gov/cder/biologics/review/etanimm052799r2.pdf.

Vance Berger and Costas Christophi [ ] offer the following guidelines for treatment allocation:

- Generate the allocation sequence in advance of screening any patients.
- Conceal the sequence from the experimenter.
- Require the experimenter to enroll all eligible subjects in the order in which they are screened.
- Verify that the subject actually received the assigned treatment.
- Conceal the proportions that have already been allocated [Schulz, 1996].
- Do not permit enrollment discretion when randomization may be triggered by some earlier response pattern.
- Conceal treatment codes until all patients have been randomized and the database is locked.

Berger [2007] notes that in unmasked trials (which are common when complementary and alternative medicines are studied), "the primary threat to allocation concealment is not the direct observation, but rather the prediction of future allocations based on the patterns in the allocation sequence that are created by the restrictions used on the randomization process."

DON'T DO THIS AT WORK

It doesn't pay to be too complicated. The randomization plan for a crossover design was generated in permuted blocks of 18. The 18 sequences were assigned with equal probabilities in the sense that, *a priori*, none of the sequences had a higher likelihood of getting assigned to a particular patient than any other. Thus, some blocks might have three instances of the first treatment sequence, none of the second, one of the third, and so forth.

The drugs were provided in sealed packets, so that with the complex treatment allocation scheme described above, investigators were unlikely to guess what treatment sequence a patient would be receiving. But the resultant design was so grossly unbalanced that period and treatment effects were confounded.

An appropriate treatment allocation scheme would have provided for the 18 treatment sequences to be allocated in random order, the order varying from block to block.

Blocked Randomization, Restricted Randomization, and Adaptive Designs

All the above caveats apply to these procedures as well. The use of an advanced statistical technique does not absolve its users of the need to exercise common sense. Observers must be kept blinded to the treatment received.

Don't Be Too Clever

Factorial experiments make perfect sense when employed by chemical engineers, as do the Greco-Latin squares used by agronomists. But social scientists should avoid employing them in areas which are less well understood than chemistry and agriculture.

Fukada [1993] reported, "Fifteen female rats were divided into three groups at the age of 12 months. Ten rats were ovariectimized and five of them were fed a diet containing 1% Ca as AAACa and the other five rats were fed a low CA diet containing 0.03% calcium. Remaining five rats were fed a Control Diet containing 1% Ca as $CaCO_3$ as the Control Group." Putting this description into an experimental design matrix yields the following nonsensical result:

	Diet		
Surgery	A	B	C
Yes	X	X	
No			X

ARE EXPERIMENTS REALLY NECESSARY?

In the case of rare diseases and other rare events, it is tempting to begin with the data in hand, that is, the records of individuals known to have the disease, rather than to draw a random and expensive sample from the population at large. There is a right way and a wrong way to conduct such studies.

The wrong way is to reason backward from effect to cause. Suppose that the majority of patients with pancreatic cancer are coffee drinkers. Does this mean that coffee causes pancreatic cancer? Not if the majority of individuals in the population in which the cases occurred are also coffee drinkers.

To be sure, suppose that we create a set of *case controls*, matching each individual in the pancreatic database with an individual in the population at large of identical race, sex, and age and with as many other nearly matching characteristics as the existing data can provide. We could then compare the incidence of coffee drinkers in the cancer database with the incidence in the matching group of case controls.

TO LEARN MORE

Good [2006] provides a series of anecdotes concerning the mythical Bumbling Pharmaceutical and Device Company that amply illustrate the results of inadequate planning. See also Andersen [1990] and Elwood [1998]. The opening chapters of Good [2001] contain numerous examples of courtroom challenges based on misleading or inappropriate samples. See also Copas and Li [1997] and the subsequent discussion.

Definitions and a further discussion of the interrelation among power and significance level may be found in Lehmann [1986], Casella and Berger [1990], and Good [2001]. You will also find discussions of optimal statistical procedures and their assumptions.

Lachin [1998], Lindley [1997], and Linnet [1999] offer guidance on sample size determination. Shuster [1993] provides sample size guidelines for clinical trials. A detailed analysis of bootstrap methodology is provided in Chapters 5 and 7.

Rosenberger and Lachin [2002] and Schulz and Grimes [2002] discuss randomization and blinding in clinical trials.

Recent developments in sequential design include *group sequential designs*, which involve testing not after every observation as in a fully sequential design, but rather after groups of observations, for example, after every 6 months in a clinical trial. The design and analysis of such experiments is best done using specialized software such as S + SeqTrial, which can be obtained from http://www.insightful.com.

For further insight into the principles of experimental design, light on math and complex formula but rich in insight are the lessons of the masters: Fisher [1925, 1935] and Neyman [1952]. If formulas are what you desire, see Hurlbert [1984], Jennison and Turnbull [1999], Lachin [1998], Lindley [1997], Linnet [1999], Montgomery and Myers [1995], Rosenbaum [2002], Thompson and Seber [1996], and Toutenberg [2002].

Among the many excellent texts on survey design are Fink and Kosecoff [1998], Rea, Parker, and Shrader [1997], and Cochran [1977]. For tips on formulating survey questions, see Converse and Presser [1986], Fowler and Fowler [1995] and Schroeder [1987]. For tips on improving the response rate, see Bly [1990, 1996].

PART II

STATISTICAL ANALYSIS

4

DATA QUALITY ASSESSMENT

Just as 95% of research efforts are devoted to data collection, 95% of the time remaining should be spent ensuring that the data collected warrant analysis.

A decade ago, Dr. Good found himself engaged by a man he had met at a flea market to consult for a start-up firm. The pay was generous, but it was conditional on the firm's receiving start-up capital.

Laboring on a part-time basis for six months, Dr. Good was concerned throughout both by the conditional nature of the pay and by the tentative manner in which the data were doled out to him. "May I see the raw data?" he kept asking, but each time was told by his sponsor that such a review was unnecessary.

One day, the president of the firm, heretofore glimpsed only from a distance, called Good in and asked for a summary of the results. Good responded with a renewed request for the raw data.

With marked reluctance, and only after nearly a week of stalling, the raw data were put into Good's hands. His first act was to run SAS Proc Means, which displays the mean, minimum, and maximum of each variable among other descriptive statistics.

"Is zero a reasonable outcome?" he asked one of the domain experts the next day. "Can't happen," he was told.

Good began to scan the data, searching for the entries with zero values. To his astonishment, more than half of the entries consisted of nothing beyond a name, an address (presumably fake), and a string of zeros. The emperor was naked; the claims of a substantial patient database were contrived; the executive in charge was

Common Errors in Statistics (and How to Avoid Them), Third Edition. Edited by P. I. Good and J. W. Hardin
Copyright © 2009 John Wiley & Sons, Inc.

faking the data whenever he wasn't feeding his addiction to methamphetamine. Or visiting flea markets. Dr. Good's last step was to try and collect his pay.[1] Your first step after the data are in hand must always be to run a data quality assessment (DQA). The focus of this chapter is on the tools you'll need.

OBJECTIVES

A DQA has many objectives. The first is immediate. You need to determine whether a decision or an estimate can be made with the desired level of certainty, given the quality of the data. The remaining objectives look toward future efforts:

- Were the response variables appropriate? Should additional data have been recorded?
- Were the measuring devices adequate?
- How well did the sampling design perform?
- If the same sampling design strategy were used again in a similar study, would the data support the same intended use with the desired level of certainty?
- Were sufficient samples taken (after correcting for missing data) to detect an effect of practical significance if one were present?

REVIEW THE SAMPLING DESIGN

1. Were the baseline values of the various treatment groups comparable? The baseline values (age and bone density) of the various groups studied by Fujita et al. [2000] were quite different, casting doubt on the reported findings.
2. Were the controls appropriate? The data quality assessment performed by Kelly et al. [1998] focused on the measured concentrations of various radioactive constituents in soil and creek sediments. At issue was whether radiation from the sediments exceeded background levels. Figure 4.1, taken from Figure 7 of their report, "shows that a comparison of background and site data for plutonium raises questions about the appropriateness of the plutonium background data, since the site data had lower levels than the background data."
3. Was the blinding effective? A subsample of those completing printed and online surveys should be contacted for personal interviews to verify their responses (see, for example, Nunes, Pretzlik, and Ilicak, 2005).

A subsample of nonresponders should also be contacted.

---

[1]"I can pay you now," the manager said, "but if I do, you'll never work for me again." Needless to say, Dr. Good took the money and ran to the bank.

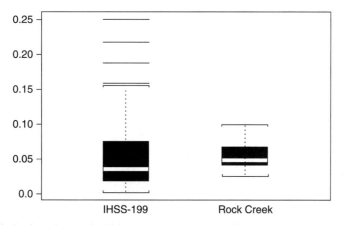

Figure 4.1. In these box-and-whisker plots taken from Kelly et al. [1998], the site data had lower levels of plutonium than the background data.

Berger [2005] is skeptical about contacting investigators. "If there is selection bias, and we ask those investigators who caused it which treatments they think each patient received, then we are essentially asking them to confess." He offers two alternatives:

1. "[U]se the randomized response technique."
2. "[S]tudy the responses of an investigator (especially if randomized response is used) to see if these responses follow the pattern mandated by the restrictions on the randomization, and to see, for example, if there are more correct guesses at the end of blocks than at the beginning of blocks." The latter would suggest that the investigator might have formed an opinion about a treatment received before it was even administered, based on prior allocations and knowledge of the restrictions on the randomization.

DATA REVIEW

During the course of the data review, inspect the database in its entirety, and generate a series of statistics and graphs.

1. Review quality assurance reports. Follow up on any discrepancies.
2. Calculate the minimum and maximum of all variables and compare them against predetermined ranges. (Ideally, this would have been done at the time the data were collected.) Generate box and whisker plots with the same goal in mind.
3. Eliminate duplicates from the database.

4. Verify that data are recorded in correct physical units and that calibration and dilution factors have been applied.

5. Characterize missing data. Problems arise in either of the following cases:
   - When the frequency of missing data is associated with the specific treatment or process that was employed.
   - When specific demographic(s) fail to complete or return survey forms, so that the remaining sample is no longer representative of the population as a whole.

6. For each variable
   - Compute a serial correlation to confirm that the observations are independent of one another.
   - Create a four-plot as described in the next section.

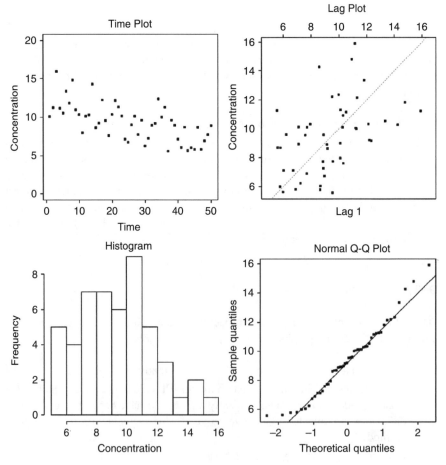

Figure 4.2. Example of a four-plot.

OUTLIERS

Outliers, extreme values, either small or large, that are well separated from the main set of observations, are frequently detected during a DQA, as they are easily spotted on a dot chart or a box whiskers plot. But as they are not signs of poor data, they should not be eliminated from the database. Rather, they should be dealt with during the subsequent analyses.

THE FOUR-PLOT

Four assumptions underlie almost all measurement processes: the data should be (1) random, (2) from a single fixed distribution, with (3) a fixed location and (4) a fixed variance. To verify these assumptions, use a four-plot consisting of a time plot, a lag plot, a histogram, and a normal probability plot.

- The data are random if the lag plot is structureless.
- If the time plot is flat and nondrifting, the data have a fixed location.
- If the time plot has a constant spread, the data have a fixed variance.
- If the histogram has multiple modes, the data may have come from multiple distributions and further stratification should be considered.

In Figure 4.2, note that the data are not quite normal (deviations from the straight line on the plot), do not have a fixed location (a downward trend in the time plot), and possibly have serial correlation present (the tendency of the lag plot to increase from left to present (the tendency of the lag plot to be increasing from left to right).

TO LEARN MORE

Consult the excellent documents available from the United States Environmental Protection Agency at http://www.epa.gov/quality/dqa.html. See, also, Husted et al. [2000].

5

ESTIMATION

Accurate, reliable estimates are essential to effective decision making. In this chapter, we review preventive measures and list the properties to look for in an estimation method. Several robust semiparametric estimators are considered, along with one method of interval estimation, the bootstrap.

PREVENTION

The vast majority of errors in estimation stem from a failure to measure what was wanted or what was intended. Misleading definitions, inaccurate measurements, errors in recording and transcription, and confounding variables plague results.

To forestall such errors, review your data collection protocols and procedure manuals before you begin, run several preliminary trials, record potentially confounding variables, monitor data collection, and review the data as they are collected.

DESIRABLE AND NOT-SO-DESIRABLE ESTIMATORS

> The method of maximum likelihood is, by far the most popular technique for deriving estimators.
>
> —Casella and Berger [1990, p. 289]

Common Errors in Statistics (and How to Avoid Them), Third Edition. Edited by P. I. Good and J. W. Hardin
Copyright © 2009 John Wiley & Sons, Inc.

The proper starting point for the selection of the "best" method of estimation is the objectives of the study: What is the purpose of our estimate? If our estimate is θ^* and the actual value of the unknown parameter is θ, what losses will we be subject to? It is difficult to understand the popularity of the method of maximum likelihood and other estimation procedures that do not take these losses into consideration.

The majority of losses will be monotone nondecreasing in nature; that is, the further apart the estimate θ^* and the true value θ, the larger our losses are likely to be. Typical forms of the loss function are the absolute deviation $|\theta^* - \theta|$, the square deviation $(\theta^* - \theta)^2$, and the jump, that is, no loss if $|\theta^* - \theta| < \delta$ and a big loss otherwise. Or the loss function may resemble the square deviation but take the form of a step function increasing in discrete increments.

Desirable estimators share the following properties: They are impartial, consistent, efficient, robust, and involve minimum loss.

Impartiality

Estimation methods should be impartial. Decisions should not depend on the accidental and quite irrelevant labeling of the samples. Nor should decisions depend on the units in which the measurements are made.

Suppose that we have collected data from two samples with the object of estimating the difference in location of the two populations involved. Suppose further that the first sample includes the values a, b, c, d, and e, the second sample includes the values f, g, h, i, j, and k, and our estimate of the difference is θ^*. If the observations are completely reversed, that is, if the first sample includes the values f, g, h, i, j, and k and the second sample includes the values a, b, c, d, and e, our estimation procedure should declare the difference to be $-\theta^*$.

The units we use in our observations should not affect the resulting estimates. We should be able to take a set of measurements in feet, convert to inches, make our estimate, convert back to feet, and get the same result as if we'd worked in feet throughout. Similarly, where we locate the zero point of our scale should not affect the conclusions.

Finally, if our observations are independent of the time of day, the season, and the day on which they were recorded (facts which ought to be verified before proceeding further), then our estimators should be independent of the order in which the observations were collected.

Consistency

Estimators should be *consistent*, that is, the larger the sample, the greater the probability that the resultant estimate will be close to the true population value.

Efficiency

One consistent estimator certainly is to be preferred to another if the first consistent estimator can provide the same degree of accuracy with fewer observations. To simplify comparisons, most statisticians focus on the *asymptotic relative efficiency* (ARE), defined as the limit with increasing sample size of the ratio of the number of observations required for each of two consistent statistical procedures to achieve the same degree of accuracy.

Robustness

Estimators that are perfectly satisfactory for use with symmetric, normally distributed populations may not be as desirable when the data come from nonsymmetric or heavy-tailed populations or when there is a substantial risk of contamination with extreme values.

When estimating measures of central location, one way to create a more robust estimator is to trim the sample of its minimum and maximum values (the procedure used when judging ice skating or gymnastics). As information is thrown away, trimmed estimators are less efficient.

In many instances, LAD (least absolute deviation) estimators are more robust than their Ordinary Least Squares (OLS) counterparts.[1]

Many *semiparametric estimators* are not only robust but provide for high ARE with respect to their parametric counterparts.

As an example of a semiparametric estimator, suppose the $\{X_i\}$ are independent and identically distributed (i.i.d.) observations with distribution $\Pr\{X_i \leq x\} = F[y - \Delta]$ and we want to estimate the location parameter Δ without having to specify the form of the distribution F. If F is normal and the loss function is proportional to the square of the estimation error, then the arithmetic mean is optimal for estimating Δ. Suppose, on the other hand, that F is symmetric but more likely to include very large or very small values than a normal distribution. Whether the loss function is proportional to the absolute value or the square of the estimation error, the median, a semiparametric estimator, is to be preferred. The median has an ARE relative to the mean that ranges from 0.64 (if the observations really do come from a normal distribution) to values well in excess of 1 for distributions with higher proportions of very large and very small values [Lehmann, 1998, p. 242]. Still, if the unknown distribution were "almost" normal, the mean would be far preferable.

If we are uncertain whether F is symmetric, then our best choice is the Hodges–Lehmann estimator, defined as the median of the pairwise averages

$$\hat{\Delta} = \text{median}_{i \leq j}(X_j + X_i)/2$$

Its ARE relative to the mean is 0.97 when F is a normal distribution [Lehmann, 1998, p. 246]. With little to lose with respect to the sample mean if F is near normal and much to gain if F is not, the Hodges–Lehmann estimator is recommended.

Suppose m observations $\{X_i\}$ and n observations $\{Y_j\}$ are i.i.d. with distributions $\Pr\{X_i \leq x\} = F[x]$ and $\Pr\{Y_j \leq y\} = F[y - \Delta]$ and we want to estimate the shift

---

[1]See, for example, Yoo [2001].

parameter Δ without having to specify the form of the distribution F. For a normal distribution F, the optimal estimator with least square losses is

$$\bar{\Delta} = \frac{1}{mn}\sum_i\sum_j(Y_j - X_i) = \bar{Y} - \bar{X}$$

the arithmetic average of the mn differences $Y_j - X_i$. Means are highly dependent on extreme values; a more robust estimator is given by

$$\hat{\Delta} = \text{median}_{ij}(Y_j - X_i)$$

Minimum Loss

The accuracy of an estimate, that is, the degree to which it comes close to the true value of the estimated parameter, and the associated losses will vary from sample to sample. A *minimum loss estimator* is one that minimizes the losses when the losses are averaged over the set of all possible samples. Thus, its form depends upon all of the following: the loss function, the population from which the sample is drawn, and the population characteristic that is being estimated. An estimate that is optimal in one situation may only exacerbate losses in another.

Minimum loss estimators in the case of least-square losses are widely and well documented for a wide variety of cases. Linear regression with an LAD loss function is discussed in Chapter 12.

Mini-Max Estimators

It's easy to envision situations in which we are less concerned with the average loss than with the maximum possible loss we may incur by using a particular estimation procedure. An estimate that minimizes the maximum possible loss is termed a mini-max estimator. Alas, few off-the-shelf mini-max solutions are available for practical cases, but see Pilz [1991] and Pinelis [1988].

Other Estimation Criteria

The expected value of an *unbiased* estimator is the population characteristic being estimated. Thus, unbiased estimators are also consistent estimators.

Minimum variance estimators provide relatively consistent results from sample to sample. While minimum variance is desirable, it may be of practical value only if the estimator is also *unbiased*. For example, 6 is a minimum variance estimator, but it offers few other advantages.

A *plug-in estimator* substitutes the sample statistic for the population statistic—for example, the sample mean for the population mean or the sample's 20th percentile for the population's 20th percentile. Plug-in estimators are consistent, but they are not always unbiased or minimum loss.

Always choose an estimator that will minimize losses.

Myth of Maximum Likelihood

The popularity of the maximum likelihood estimator is hard to comprehend other than as a vehicle whereby an instructor can demonstrate knowledge of the calculus. This estimator may be completely unrelated to the loss function and has as its sole justification that it corresponds to that value of the parameter that makes the observations most probable—provided, that is, they are drawn from a specific predetermined (and *unknown*) distribution. The observations might have resulted from a thousand other *a priori* possibilities.

A common and lamentable fallacy is that the maximum likelihood estimator has many desirable properties—that it is unbiased and minimizes the mean squared error. But this is true only for the maximum likelihood estimator of the mean of a normal distribution.[2]

Statistics instructors would be well advised to avoid introducing maximum likelihood estimation and to focus instead on methods for obtaining minimum loss estimators for a wide variety of loss functions.

INTERVAL ESTIMATES

Point estimates are seldom satisfactory in and of themselves. First, if the observations are continuous, the probability is zero that a point estimate will be correct and will equal the estimated parameter. Second, we still require some estimate of the precision of the point estimate.

In this section, we consider one form of the *interval estimate* derived from bootstrap measures of precision. A second form, derived from tests of hypotheses, will be considered in the next chapter.

A common error is to create a confidence interval in the form

$$(\text{estimate} - k * \text{standard error}, \text{estimate} + k * \text{standard error})$$

This form is applicable only when an interval estimate is desired for the mean of a normally distributed random variable. Even then, k should be determined from tables of the Student's-t distribution and not from tables of the normal distribution.

Nonparametric Bootstrap

The bootstrap can help us obtain an interval estimate for any aspect of a distribution— a median, a variance, a percentile, or a correlation coefficient—*if* the observations are independent and all come from distributions with the same value of the parameter to be estimated. This interval provides us with an estimate of the precision of the corresponding point estimate.

From the original sample, we draw a random sample (with replacement); this random sample is called a bootstrap sample. The random sample is the same size as

---

[2]It is also true in some cases for very large samples. How large the sample must be in each case will depend both upon the parameter being estimated and upon the distribution from which the observations are drawn.

the original sample and is used to compute the sample statistic. We repeat this process a number of times, 1000 or so, always drawing samples with replacement from the original sample. The collection of computed statistics for the bootstrap samples serves as an empirical distribution of the sample statistic of interest to which we compare the value of the sample statistic computed from the original sample.

For example, here are the heights of a group of 22 adolescents, measured in centimeters and ordered from shortest to tallest:

137.0   138.5   140.0   141.0   142.0   143.5   145.0   147.0   148.5

150.0   153.0   154.0   155.0   156.5   157.0   158.0   158.5   159.0

160.5   161.0   162.0   167.5

The median height lies somewhere between 153 and 154 centimeters. If we want to extend this result to the population, we need an estimate of the precision of this average.

Our first bootstrap sample, arranged in increasing order of magnitude for ease in reading, might look like this:

138.5   138.5   140.0   141.0   141.0   143.5   145.0   147.0   148.5

150.0   153.0   154.0   155.0   156.5   157.0   158.5   159.0   159.0

159.0   160.5   161.0   162.0

Several of the values have been repeated. This is not surprising, as we are sampling with replacement, treating the original sample as a stand-in for the much larger population from which the original sample was drawn. The minimum of this bootstrap sample is 138.5, higher than that of the original sample; the maximum at 162.0 is less than the original, while the median remains unchanged at 153.5.

137.0   138.5   138.5   141.0   141.0   142.0   143.5   145.0   145.0

147.0   148.5   148.5   150.0   150.0   153.0   155.0   158.0   158.5

160.5   160.5   161.0   167.5

In this second bootstrap sample, again we find repeated values; this time the minimum, maximum, and median are 137.0, 167.5, and 148.5, respectively.

The medians of 50 bootstrapped samples drawn from our sample ranged between 142.25 and 158.25, with a median of 152.75 (see Fig. 5.1). These numbers provide insight into what might have been had we sampled repeatedly from the original population.

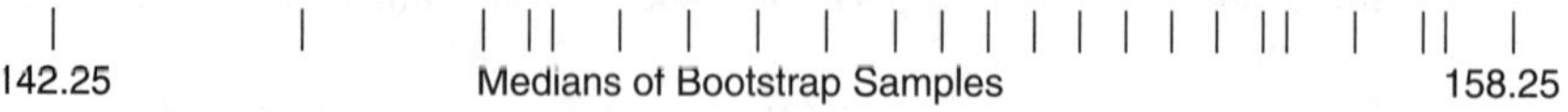

Figure 5.1. Scatterplot of 50 bootstrap medians derived from a sample of heights.

We can improve on the interval estimate {142.25, 158.25} if we are willing to accept a small probability that the interval will fail to include the true value of the population median. We will take several hundred bootstrap samples instead of a mere 50, and use the 5th and 95th percentiles of the resulting bootstrap (empirical) distribution to establish the boundaries of a 90% confidence interval.

This method might be used equally well to obtain an interval estimate for any other population attribute: the mean and variance, the 5th percentile or the 25th, and the interquartile range. When several observations are made simultaneously on each subject, the bootstrap can be used to estimate covariances and correlations among the variables. The bootstrap is particularly valuable when trying to obtain an interval estimate for a ratio or for the mean and variance of a nonsymmetric distribution.

Unfortunately, such intervals have two deficiencies:

1. They are biased, that is, they are more likely to contain certain false values of the parameter being estimated than the true value [Efron, 1987].
2. They are wider and less efficient than they could be [Efron, 1987], that is, they *conservatively* fail to establish significance when such significance exists.

Two methods have been proposed to correct these deficiencies; let us consider each in turn.

The first is the Hall–Wilson [1991] corrections, in which the bootstrap estimate is Studentized. For the one-sample case, we want an interval estimate based on the distribution of $(\hat{\theta}_b - \hat{\theta})/s_b$, where $\hat{\theta}$ and $\hat{\theta}_b$ are the estimates of the unknown parameter based on the original and bootstrap sample, respectively, and s_b denotes the standard deviation of the bootstrap sample. An estimate $\hat{\sigma}$ of the population variance is required to transform the resultant interval into one about θ [see Carpenter and Bithell, 2000].

For the two-sample case, we want a confidence interval based on the distribution of

$$\frac{\hat{\theta}_{nb} - \hat{\theta}_{mb}}{\sqrt{\dfrac{(n-1)s_{nb}^2 + (m-1)s_{mb}^2}{n+m-2}\,(1/n + 1/m)}},$$

where n, m, and s_{nb}, s_{mb} denote the sample sizes and standard deviations, respectively, of the bootstrap samples. Applying the Hall–Wilson corrections, we obtain narrower interval estimates. Even though this interval estimate is narrower, it is still *more* likely to contain the true value of the unknown parameter.

The bias-corrected and accelerated BC_a interval due to Efron and Tibshirani [1986] also represents a substantial improvement, though for samples under size 30, the properties of the interval are still suspect. The idea behind these intervals comes from the observation that percentile bootstrap intervals are most accurate when the estimate is symmetrically distributed about the true value of the parameter and the tails of the estimate's distribution drop off rapidly to zero. The symmetric, bell-shaped normal distribution depicted in Figure 5.2 represents this ideal.

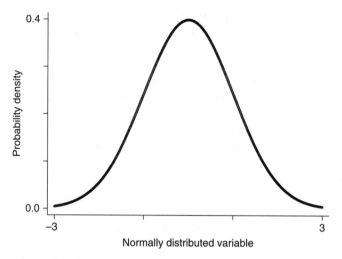

Figure 5.2. Bell-shaped symmetric curve of a normal distribution.

Suppose θ is the parameter we are trying to estimate, $\hat{\theta}$ is the estimate, and we establish a monotone increasing transformation m such that $m(\theta)$ is normally distributed about $m(\hat{\theta})$. We could use this normal distribution to obtain an unbiased confidence interval and then apply a back transformation to obtain an almost unbiased confidence interval.[3] That we discovered and implemented a monotone transformation is what allows us to invert that function to transform the interval based on normality back to the original (possibly asymmetric and platykurtotic) distribution. Even with these modifications, we do not recommend the use of the nonparametric bootstrap with samples of fewer than 100 observations. Simulation studies suggest that with small sample sizes, the coverage is far from exact and the endpoints of the intervals vary widely from one set of bootstrap samples to the next. For example, Tu and Zhang [1992] report that with samples of size 50 taken from a normal distribution, the actual coverage of an interval estimate rated at 90% using the BC$_a$ bootstrap is 88%. When the samples are taken from a mixture of two normal distributions (a not uncommon situation with real-life data sets), the actual coverage is 86%. With samples of only 20 in number, the actual coverage is only 80%.

More serious than the disappointing coverage probabilities discussed is that the endpoints of the resulting interval estimates from the bootstrap may vary widely from one set of bootstrap samples to the next. For example, when Tu and Zhang drew samples of size 50 from a mixture of normal distributions, the average of the left limit of 1000 bootstrap samples taken from each of 1000 simulated data sets was 0.72 with a standard deviation of 0.16; the average and standard deviation of the right limit were 1.37 and 0.30, respectively.

---

[3]Stata^{TM} provides for bias-corrected intervals via its -bstrap command. R and S-Plus both include BC$_a$ functions. A SAS macro is available at http://cuke.hort.ncsu.edu/cucurbit/wehner/software/pathsas/jackboot.txt.

Parametric Bootstrap

Even when we know the form of the population distribution, the use of the *parametric bootstrap* to obtain interval estimates may prove advantageous either because the parametric bootstrap provides more accurate answers than textbook formulas or because no textbook formulas exist.

Suppose we know the observations come from a normal distribution and we want an interval estimate for the standard deviation. We would draw repeated bootstrap samples from a normal distribution the mean of which is the sample mean and the variance of which is the sample variance. As a practical matter, we would draw an element from a $N(0,1)$ population, multiply by the sample standard deviation, and then add the sample mean to obtain an element of our bootstrap sample. By computing the standard deviation of each bootstrap sample, an interval estimate for the standard deviation of the population may be constructed from the collection of statistics.

IMPROVED RESULTS

In many instances, we can obtain narrower interval estimates that have a greater probability of including the true value of the parameter by focusing on sufficient statistics, pivotal statistics, and admissible statistics.

A statistic T is *sufficient* for a parameter if the conditional distribution of the observations given this statistic T is independent of the parameter. If the observations in a sample are exchangeable, then the order statistics of the sample are sufficient; that is, if we know the order statistics $x_{(1)} \leq x_{(2)} \leq \cdots \leq x_{(n)}$, then we know as much about the unknown population distribution as we would if we had the original sample in hand. If the observations are on successive independent binomial trials that result in either success or failure, then the number of successes is sufficient to estimate the probability of success. The minimal sufficient statistic that reduces the observations to the fewest number of discrete values is always preferred.

A *pivotal* quantity is any function of the observations and the unknown parameter that has a probability distribution that does not depend on the parameter. The classic example is the Student's t whose distribution does not depend on the population mean or variance when the observations come from a normal distribution.

A decision procedure d based on a statistic T is *admissible* with respect to a given loss function L, provided there does not exist a second procedure d^* whose use would result in smaller losses whatever the unknown population distribution.

The importance of admissible procedures is illustrated in an expected way by Stein's paradox. The sample mean, which plays an invaluable role as an estimator of the population mean of a normal distribution for a single set of observations, proves to be inadmissible as an estimator when we have three or more independent sets of observations to work with. Specifically, if $\{X_{ij}\}$ are independent observations taken from four or more distinct normal distributions with means θ_i and variance 1, and losses are proportional to the square of the estimation error, then the estimators $\hat{\theta}_i = \bar{X}_{..} + (1 - [k-3]/S^2)(\bar{X}_{i.} - \bar{X}_{..})$, where $S^2 = \sum_{i=1}^{k}(\bar{X}_{i.} - \bar{X}_{..})^2$, have smaller expected losses than the individual sample means, regardless of the actual values of the population means [see Efron and Morris, 1977].

SUMMARY

Desirable estimators are impartial, consistent, efficient, robust, and minimum loss. Interval estimates are to be preferred to point estimates; they are less open to challenge, for they convey information about the estimate's precision.

TO LEARN MORE

Selecting more informative endpoints is the focus of Berger [2002] and Bland and Altman [1995].

Lehmann and Casella [1998] provide a detailed theory of point estimation.

Robust estimators are considered in Huber [1981], Maritz [1996], and Bickel et al. [1993]. Additional examples of both parametric and nonparametric bootstrap estimation procedures may be found in Efron and Tibshirani [1993]. Shao and Tu [1995, Section 4.4] provide a more extensive review of bootstrap estimation methods along with a summary of empirical comparisons.

Carroll and Ruppert [2000] show how to account for differences in variances between populations; this is a necessary step if one wants to take advantage of Stein–James–Efron–Morris estimators.

Bayes estimators are considered in Chapter 7.

6

TESTING HYPOTHESES: CHOOSING A TEST STATISTIC

Forget "large-sample" methods. In the real world of experiments samples are so nearly always "small" that it is not worth making any distinction, and small-sample methods are no harder to apply.

—George Dyke [1997].

Statistical tests should be chosen before the data are analyzed, and the choice should be based on the study design and distribution of the data, not the results.

—Cara H. Olsen

Life constantly forces us to make decisions. If life weren't so uncertain, the correct choice would always be obvious. But life isn't certain, and the choice isn't obvious. As always, proper application of statistical methods can help us to cope with uncertainty, but it cannot eliminate it.

In the preceding chapter on estimation, our decision consisted of choosing one value or one interval out of an unlimited number of possibilities. Each decision had associated with it a potential loss, which increased as the difference between the correct decision and our decision increased.

In this chapter on hypothesis testing, our choices reduce to three possibilities:

1. To embrace or accept a primary hypothesis.
2. To reject the primary hypothesis and embrace or accept one or more alternative hypotheses.
3. To forgo making a decision until we have gathered more data.

Among the most common errors in (prematurely) published work is the failure to recognize that the last decision listed above is the correct one.

Common Errors in Statistics (and How to Avoid Them), Third Edition. Edited by P. I. Good and J. W. Hardin
Copyright © 2009 John Wiley & Sons, Inc.

FIRST STEPS

Before we can apply statistical methods properly, we need to establish all of the following:

1. The primary hypothesis and the alternative hypotheses of interest.
2. The nature and relative magnitude of the losses associated with erroneous decisions.
3. The type of data that is to be analyzed.
4. The statistical test that will be employed.
5. The significance level of each test that is to be performed.

Moreover, all of these steps must be completed *before* the data are examined.

The first step allows us to select a testing procedure that maximizes the probability of detecting such alternatives. For example, if our primary hypothesis in a k-sample comparison is that the means of the k-populations from which the samples are taken are the same, and the alternative is that we anticipate an ordered dose response, then the optimal test will be based on the correlation between the doses and the responses, and *not* the F-ratio of the between-sample and within-sample variances.

If we fail to complete step 2, we also risk selecting a less powerful statistic. Suppose, once again, we are making a k-sample comparison of means. If our anticipated losses are proportional to the squares of the differences among the population means, then our test should be based on the F-ratio of the between-sample and within-sample variances. But if our anticipated losses are proportional to the absolute values of the differences among the population means, then our test should be based on the ratio of the between-sample and within-sample absolute deviations.

Several commercially available statistics packages automatically compute the p-values associated with several tests of the same hypothesis—for example, that of the Wilcoxon and the t-test. Rules 3 and 4 state the obvious. Rule 3 reminds us that the type of test to be employed will depend upon the type of data to be analyzed—binomial trials, categorical data, ordinal data, measurements, and time to events. Rule 4 reminds us that we are not free to choose the p-value that best fits our preconceptions but must specify the test we employ *before* we look at the results.

Collectively, Rules 1 through 5 dictate that we need always specify whether a test will be one-sided or two-sided *before* a test is performed and before the data are examined. Two notable contradictions of this collection of rules arose in interesting court cases.

In the first of these, the Commissioner of Food and Drugs had terminated provisional approval of a food coloring, Red No. 2, and the Certified Color Manufacturers sued.[1]

Included in the data submitted to the court was Table 6.1a; an analysis of this table by Fisher's Exact Test reveals a statistically significant dose response to the dye.[1] The

---

[1] *Certified Color Manufacturers Association v. Mathews*, 543 F.2d 284 (1976 DC), Note 31.

TABLE 6.1a. Rats Fed Red No. 2

	Low Dose	High Dose
No cancer	14	14
Cancer	0	7

TABLE 6.1b. Rats Fed Red No. 2

	Low Dose	High Dose
No cancer	7	21
Cancer	7	0

TABLE 6.2. Scores on Department Examinations[3]

	Caucasians		African-Americans		
	#	Range	#	Range	Cutoff
Assistant fire chief	25	73–107	2	71–99	100
Fire deputy chief	45	76–106	1	97	100
Fire battalion chief	99	58–107	6	83–93	94

response is significant, that is, if the court tests the null hypothesis that Red No. 2 does not affect the incidence of cancer against the one-sided alternative that high doses of Red No. 2 do induce cancer, at least in rats. The null hypothesis is rejected because only a small fraction of the tables with the marginals shown in Table 6.1a reveal a toxic effect as extreme as the one actually observed.

This is an example of a one-tailed test. Should it have been? What would your reaction have been if the results had taken the form shown in Table 6.1b, that is, that Red No. 2 prevented tumors, at least in rats?

Should the court have guarded against this eventuality, that is, should they have performed a two-tailed test that would have rejected the null hypothesis if either extreme were observed? Probably not, but a Pennsylvania federal district court was misled into making just such a decision in *Commonwealth of Pennsylvania et al. v. Rizzo* et al.[23]

In the second illuminating example, African-American firemen sued the city of Philadelphia. The city's procedures for determining which firemen would be promoted included a test that was alleged to be discriminatory against African-Americans. The results of the city's promotion test are summarized in Table 6.2.

Given that the cutoff point always seems to be just above the African-American candidates' highest score, these results look suspicious. Fisher's Exact Test applied to the pass/fail results was only marginally significant at .0513; still, the court ruled

---

[2]466 F. Supp 1219 (E.D. PA 1979).
[3]Data abstracted from Appendix A of 466 F. Supp 1219 (E.D. PA 1979).

that "we will not reject the result of plaintiffs' study simply by mechanically lining it up with the 5% level."[4] Do you agree with this reasoning? We do.

The plaintiffs argued for the application of a one-tailed test ("Does a smaller proportion of African-Americans score at or above the cutoff?"), but the defendants insisted that a two-tailed test is the correct comparison ("Are there differences in the proportions of African-American and Caucasian candidates scoring at or above the cutoff point?"). The court agreed—in error, we feel, given the history of discrimination against African-Americans—to consider the two-tailed test as well as the one-tailed one [see Good, 2001, Section 9.1].

TEST ASSUMPTIONS

As noted in previous chapters, before any statistical test can be performed and a p-value or confidence interval derived, we must establish all of the following:

1. The sample was selected at random from the population or from specific subsets (strata) of the population of interest.
2. Subjects were assigned to treatments at random.
3. Observations and observers are free of bias.

To these guidelines, we now add the following:

4. All assumptions are satisfied.

Every statistical procedure relies on certain assumptions for correctness. Errors in testing hypotheses come about either because the assumptions underlying the chosen test are not satisfied or because the chosen test is less powerful than other competing procedures. We shall study each of these lapses in turn.

Virtually all statistical procedures rely on the assumption that the observations are independent.

Virtually all statistical procedures require at least one of the following successively weaker assumptions be satisfied under the null hypothesis:

1. The observations are identically distributed and their distribution is known.
2. The observations are exchangeable, that is, their joint distribution remains unchanged when the labels on the observations are exchanged.
3. The observations are drawn from populations in which a specific parameter is the same across the populations.

The first assumption is the strongest. If it is true, the following two assumptions are also true. The first assumption must be true for a parametric test to provide an exact significance level. If the second assumption is true, the third assumption is also true. The second assumption must be true for a permutation test to provide an exact significance level.

The third assumption is the weakest. It must be true for a bootstrap test to provide an exact significance level asymptotically.

---

[4]Ibid. 1228–1229.

TABLE 6.3. Types of Statistical Tests of Hypotheses

Test Type	Definition	Example
Exact	Stated significance level is exact, not approximate.	t-test when observations are i.i.d. normal; permutation test when observations are exchangeable.
Parametric	Obtains cut-off points from specific parametric distribution.	t-test.
Semiparametric bootstrap	Obtains cut-off points from percentiles of bootstrap distribution of parameter.	
Parametric bootstrap	Obtains cut-off points from percentiles of parameterized bootstrap distribution of parameter.	
Permutation	Obtains cut-off points from distribution of test statistic obtained by rearranging labels.	Tests may be based upon the original observations, on ranks, on normal or Savage scores, or on U-statistics.

An immediate consequence of the first two assumptions is that if observations come from a multiparameter distribution, then all parameters, not just the one under test, must be the same for all observations under the null hypothesis. For example, a t-test comparing the means of two populations requires the variation of the two populations to be the same.

For parametric tests and parametric bootstrap tests, under the null hypothesis the observations must all come from a distribution of a specific form.

Let us now explore the implications of these assumptions in a variety of practical testing situations, including comparing the means of two populations, comparing the variances of two populations, comparing the means of three or more populations, and testing for significance in two-factor and higher-order experimental designs.

In each instance, before we choose[5] a statistic, we check which assumptions are satisfied, which procedures are most robust to violation of these assumptions, and which are most powerful for a given significance level and sample size. To find the most powerful test, we determine which procedure requires the smallest sample size for given levels of Type I and Type II error.

BINOMIAL TRIALS

With today's high-speed desktop computers, a (computationally convenient) normal approximation is no longer an excusable shortcut when testing that the probability of success has a specific value; use binomial tables for exact, rather than approximate,

---

[5] Whether Republican or Democrat, Liberal or Conservative, male or female, we have the right to choose, and we need not be limited by what our textbook, half-remembered teacher pronouncements, or software dictate.

inference. To avoid error, if sufficient data are available, test to see that the probability of success has not changed over time or from clinical site to clinical site.

When comparing proportions, two cases arise. If $0.1 < p < 0.9$, use Fisher's Exact Test. To avoid mistakes, test for a common odds ratio if several laboratories or clinical sites are involved. This procedure is described in the StatXact manual.

If p is close to zero, as it would be with a relatively rare event, a different approach is called for [see Lehmann, 1986, pp. 151–154]. Recently, Dr. Good had the opportunity to participate in the conduct of a very large-scale clinical study of a new vaccine. He had not been part of the design team, and when he read over the protocol, he was stunned to learn that the design called for inoculating and examining 100,000 patients—50,000 with the experimental vaccine and 50,000 controls with a harmless saline solution.

Why so many? The disease at which the vaccine was aimed was relatively rare. Suppose we could expect 0.8%, or 400 of the controls, to contract the disease and 0.7%, or 350 of those vaccinated, to contract it. Put another way, if the vaccine was effective, we would expect 400 out of every 750 patients who contracted the disease to be controls. By contrast, if the vaccine was ineffective (and innocuous), we would expect 50% of the patients who contracted the disease to be controls.

In short, of the 100,000 subjects we'd exposed to a potentially harmful vaccine, only 750 would provide information to use for testing the vaccine's effectiveness.

The problem of comparing samples from two Poisson distributions boils down to testing the proportion of a single binomial. And the power of this test that started with 100,000 subjects is based on the outcomes of only 750.

But 750 was merely the expected value; it could not be guaranteed. In fact, less than 100 of those inoculated—treated and control—contracted the disease. The result was a test with extremely low power. As always, the power of a test depends not on the number of subjects with which one starts a trial but on the number with which one ends it.

CATEGORICAL DATA

The chi-square statistic that is so often employed in the analysis of contingency tables $\sum (f_{ij} - Ef_{ij})^2 / Ef_{ij}$ does *not* have the chi-square distribution. That distribution represents an asymptotic approximation of the statistic that is valid only with very large samples. To obtain exact tests of independence in a 2×2 table, use Fisher's Exact Test.

Consider Table 6.4, where we've recorded the results of a comparison of two drugs. It seems obvious that Drug B offers significant advantages over Drug A. Or does it? A chi-square analysis by parametric means in which the value of the chi-squared

TABLE 6.4. A 2 × 2 Contingency Table

	Drug A	Drug B
Response	5	9
No response	5	1

statistic is compared with a table of the chi-square distribution yields an erroneous p-value of 3%. But Fisher's Exact Test yields a one-sided p-value of only 7%. The evidence of advantage is inconclusive, and further experimentation is warranted.

As in Fisher [1935], we determine the proportion of tables with the same marginals that are as extreme as or more extreme than our original table.

The problem lies in defining what is meant by "extreme." The errors lie in failing to report how we arrived at our definition.

For example, in obtaining a two-tailed test for independence in a 2×2 contingency table, we can treat each table strictly in accordance with its probability under the multinomial distribution (Fisher's method) or weight each table by the value of the Pearson chi-square statistic for that table.

Stratified 2×2 Tables

To obtain exact tests of independence in a set of stratified 2×2 tables, first test for the equivalence of the odds ratios using the method of Mehta, Patel, and Gray [1985]. If the test for equivalence is satisfied, combine the data and use Fisher's Exact Test.

Unordered $R \times C$ Tables

In testing for differences in an $R \times C$ contingency table with unordered categories, possible test statistics include Freeman–Halton, chi-square, and the log-likelihood ratio $\sum \sum f_{ij} \log[f_{ij}f_{..}/f_{i.}f_{.j}]$. Regardless of which statistic is employed, one should calculate the exact significance levels of the test statistic by deriving its permutation distributions using the method of Mehta and Patel [1986].

The chief errors in practice lie in failing to report all of the following:

- Whether we used a one-tailed or two-tailed test and why.
- Whether the categories are ordered or unordered.
- Which statistic was employed and why.

Chapter 11 contains a discussion of a final, not inconsiderable source of error: the neglect of confounding variables that may be responsible for creating an illusory association or concealing an association that actually exists.

TIME-TO-EVENT DATA (SURVIVAL ANALYSIS)

In survival studies and reliability analyses, we follow each subject and/or experiment unit until either some event occurs or the experiment is terminated; the latter observation is referred to as *censored*. The principal sources of error are the following:

- Lack of independence within a sample.
- Lack of independence of censoring.
- Too many censored values.
- Wrong test employed.

See also, Altman et al. [1995].

Lack of Independence within a Sample

Lack of independence within a sample is often caused by the existence of an implicit factor in the data. For example, if we are measuring survival times for cancer patients, diet may be correlated with survival times. If we do not collect data on the implicit factor (diet in this case) and the implicit factor has an effect on survival times, then we no longer have a sample from a single population. Rather, we have a sample that is a mixture drawn from several populations, one for each level of the implicit factor, each with a different survival distribution.

Implicit factors can also affect censoring times by affecting the probability that a subject will be withdrawn from the study or lost to follow-up. For example, younger subjects may tend to move away (and be lost to follow-up) more frequently than older subjects, so that age (an implicit factor) is correlated with censoring. If the sample under study contains many younger people, the results of the study may be substantially biased because of the different patterns of censoring. This violates the assumption that the censored and noncensored values all come from the same survival distribution.

Stratification can be used to control for an implicit factor. For example, age groups (such as under 50, 51–60, 61–70, and 71 or older) can be used as strata to control for age. This is similar to using blocking in analysis of variance. The goal is to have each group/stratum combination's subjects have the same survival distribution.

Lack of Independence of Censoring

If the pattern of censoring is not independent of the survival times, then survival estimates may be too high (if subjects who are more ill tend to be withdrawn from the study), or too low (if subjects who will survive longer tend to drop out of the study and are lost to follow-up).

If a loss or withdrawal of one subject could increase the probability of loss or withdrawal of other subjects, this would also lead to lack of independence between censoring and the subjects.

Survival tests rely on independence between censoring times and survival times. If independence does not hold, the results may be inaccurate.

An implicit factor not accounted for by stratification may lead to a lack of independence between censoring times and observed survival times.

Many Censored Values

A study may end up with many censored values as a result of having large numbers of subjects withdrawn or lost to follow-up, or from having the study end while many subjects are still alive. Large numbers of censored values decrease the equivalent number of subjects exposed (at risk) at later times, reducing the effective sample sizes.

A high censoring rate may also indicate problems with the study: ending too soon (many subjects still alive at the end of the study) or a pattern in the censoring (many subjects withdrawn at the same time, younger patients being lost to follow-up sooner than older ones, etc.).

Survival tests perform better when the censoring is not too heavy and, in particular, when the pattern of censoring is similar across the different groups.

Which Test?

Type I Censoring. A most powerful test for use when data are censored at one end only was developed by Good [1989, 1991, 1992]. It should be employed in the following situations:

- Radioimmune assay and other assays where some observations may fall into the nonlinear portion of the scale.
- Time-to-event trials with equipment that are terminated after a fixed period.
- Time-to-event trials with animals that are terminated after a fixed period.

Type II Censoring. Kaplan–Meier survival analysis (KMSA) is the appropriate starting point, as Good's test is not appropriate for use in clinical trials for which the times are commonly censored at both ends. KMSA can estimate survival functions even in the presence of censored cases and requires minimal assumptions.

If covariates other than time are thought to be important in determining duration to outcome, results reported by KMSA will represent misleading averages obscuring important differences in groups formed by the covariates (e.g., men vs. women). Since this is often the case, methods that incorporate covariates, such as event-history models and Cox regression, may be preferred.

For small samples, the permutation distributions of the Gehan–Breslow, Mantel–Cox, and Tarone–Ware survival test statistics, and not the chi–square distribution, should be used to compute *p*-values. If the hazard or survival functions are not parallel, then none of the three tests (Gehan–Breslow, Mantel–Cox, or Tarone–Ware) will be particularly good at detecting differences between the survival functions. Before performing any of these tests, examine a Kaplan–Meier plot, plots of the life table survival functions, and plots of the life table hazard functions for each sample to see whether their graphs cross, as in Figure 6.1.

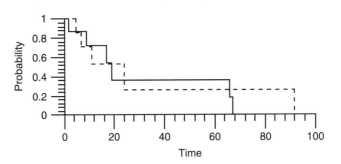

Figure 6.1. Kaplan–Meier plot showing crossing survival functions.

Comparing Treatments

Buyse and Piedbois [1996] describe four further errors that can result in misleading treatment comparisons:

1. *Comparing summary statistics from nonrandomized studies.* Regression analyses performed on summary statistics ignore the variability in the independent variable(s) and provide biased estimates of the regression slope at the individual level.

2. *Failing to match patients in the different treatment groups.* A correlation between summary statistics on response and survival may indicate merely a different patient mix in the different studies. One would expect to observe low response rates and short survival times in studies that had accrued mostly patients with far-advanced disease and in poor general condition. Conversely, one would expect to observe high response rates and long survival times in studies using patients with limited disease and in good general condition. A significant correlation between summary statistics on response and survival would, in that case, imply no causality of the relationship and provide no evidence whatsoever that if some treatment improved the response, then that treatment would also prolong survival.

3. *Ignoring the variability in the independent variable(s).* The random effects model due to Torri et al. [1992] is recommended if only summary data are available. Still, even with randomized studies, individual patient data should always be used in preference to summary statistics.

4. *Time-biased sampling.* In some cases, time bias can be eliminated by defining a "landmark period" during which patients are observed for response. Further analysis should distinguish those who survive this landmark period and those who do not.

The landmark method is adequate only when responses occur soon after starting treatment, not when responses may appear later in the course of the disease. For responses that can occur over extended periods of time, response must be considered as a time-dependent covariate.

COMPARING THE MEANS OF TWO SETS OF MEASUREMENTS

The most common test for comparing the means of two populations is based upon Student's t. For Student's t-test to provide significance levels that are exact rather than approximate, all the observations must be independent and, under the null hypothesis, all the observations must come from identical normal distributions.

Even if the distribution is not normal, the significance level of the t-test is almost exact for sample sizes greater than 12; for most of the distributions one encounters in practice,[6] the significance level of the t-test is usually within 1% or so of the correct value for sample sizes between 6 and 12.

---

[6]Here and throughout this book, we deliberately ignore the many exceptional cases, the delight of the true mathematician, that one is unlikely to encounter in the real world.

For testing against nonnormal alternatives, more powerful tests than the t-test exist. For example, a permutation test replacing the original observations with their normal scores is more powerful than the t-test [Lehmann, 1986].

Permutation tests are derived by looking at the distribution of values the test statistic would take for each of the possible assignments of treatments to subjects. For example, if in an experiment two treatments were assigned at random to six subjects, so that three subjects got one treatment and three the other, there would have been a total of 20 possible assignments of treatments to subjects.[7] To determine a p-value, we compute for the data in hand each of the 20 possible values the test statistic might have taken. We then compare the actual value of the test statistic with these 20 values. If our test statistic corresponds to the most extreme value, we say that $p = 1/20 = 0.05$ (or $1/10 = 0.10$ if this is a two-tailed permutation test).

Against specific normal alternatives, this two-sample permutation test provides a most powerful unbiased test of the distribution-free hypothesis that the centers of the two distributions are the same [Lehmann, 1986, p. 239]. For large samples, its power against normal alternatives is almost the same as that of Student's t-test [Albers, Bickel, and van Zwet, 1976]. Against other distributions, by appropriate choice of the test statistic, its power can be superior [Lambert, 1985; Maritz, 1996].

Multivariate Comparisons

A test based on several variables simultaneously, a *multivariate test*, can be more powerful than a test based on a single variable alone, provided that *the additional variables are relevant*. Adding variables that are unlikely to have value in discriminating among the alternative hypotheses simply because they are included in the data set can only result in a loss of power.

Unfortunately, what works when making a comparison between two populations based on a single variable fails when we attempt a *multivariate comparison*. Unless the data are multivariate normal, Hötelling's T^2, the multivariate analog of Student's t, will not provide tests with the desired significance level. Only samples far larger than those we are likely to afford in practice are likely to yield multivariate results that are close to multivariate normal. Still, an exact significance level can be obtained in the multivariate case regardless of the underlying distribution only by making use of the permutation distribution of Hötelling's T^2.

Suppose we had a series of multivariate observations on m control subjects and n subjects who had received a new treatment. Here's how we would construct a multivariate test for a possible treatment effect:

1. First, we would compute Hötelling's T^2 for the data at hand.
2. Next, we would take the m control labels and the n treatment labels and apply them at random to the $n + m$ vectors of observations. Listings in the R, C, and other computing languages for carrying out this step are found in Good [2006]. Note that this relabeling can be done in $m + n$; choose n or $(m + n)!/(m!n!)$ ways.

---

[7]Interested readers may want to verify this for themselves by writing out all the possible assignments of six items into two groups of three: 1 2 3/4 5 6/1 2 4/3 5 6, and so forth.

3. Then we would compute Hötelling's T^2 for the data as they are now relabeled.

4. We would then repeat Steps 2 and 3 a large number of times to obtain a permutation (empirical) distribution of possible values of Hötelling's T^2 for the data we've collected.

5. Finally, we would compare the value of Hötelling's T^2 we obtained at Step 1 with this empirical distribution. If the original value is an extreme one, lying in the tail of the permutation distribution, then we would reject the null hypothesis.

If only two or three variables are involved, a graph can sometimes be a more effective way of communicating results than a misleading *p*-value based on the

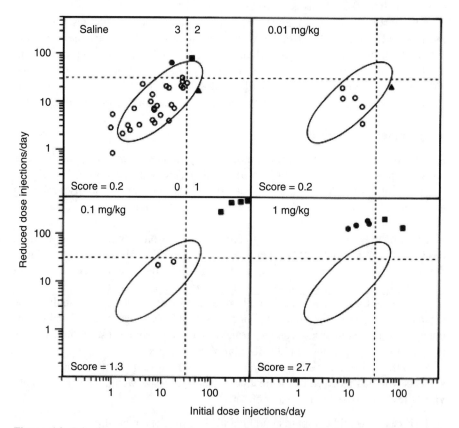

Figure 6.2. Injection rates and scores for rats self-administering saline and morphine using the pneumatic syringe and the new protocol. The ellipse is the 90% confidence limits for saline control rats based upon the assumption of a normal bivariate distribution of injection rates corresponding to the initial and reduced dose periods. The dashed lines represent the 90% confidence limits for saline self-administration for the initial and reduced doses individually. The scores for points falling in each quadrant formed by these lines are shown with the saline data. Open circles, score 0; solid triangles, score 1; solid squares, score 2; and solid circles, score 3. Note that injection rates are plotted to a logarithmic scale. *Source:* Reproduced with kind permission of Springer Science + Business Media from Weeks and Collins [1987].

parametric distribution of Hötelling's T^2. As an example, compare the graph in Weeks and Collins [1987] (Fig. 6.2) with the analysis of the same data in Collins et al. [1984].

Options

Alas, more and more individuals seem content to let their software do their thinking for them. It won't.

Your first fundamental decision is whether to do a one-tailed or a two-tailed test. If you are testing against a one-sided alternative—for example, no difference versus improvement—then you require a one-tailed or one-sided test. If you are doing a head-to-head comparison—which alternative is best?—then a two-tailed test is required.

Note that in a two-tailed test, the tails need not be equal in size, but they should be portioned out in accordance with the relative losses associated with the possible decisions [Moyé, 2000, pp. 152–157].

Second, you must decide whether your observations are paired (as would be the case when each individual serves as its own control) or unpaired and use the paired or unpaired t-test.

Testing Equivalence

When the logic of a situation calls for demonstration of similarity rather than differences among responses to various treatments, equivalence tests are often more relevant than tests with traditional no-effect null hypotheses [Anderson and Hauck, 1986; Dixon, 1998, pp. 257–301].

Two distributions, F and G, such that $G[x] = F[x - \delta]$ are said to be equivalent, provided that $|\delta| < \Delta$, where Δ is the smallest difference of clinical significance. To test for equivalence, we obtain a confidence interval for δ, rejecting equivalence *only* if this interval contains values in excess of $|\Delta|$. The width of a confidence interval decreases as the sample size increases; thus, a very large sample may be required to demonstrate equivalence, just as a very large sample may be required to demonstrate a clinically significant effect.

Operationally, establishing equivalence can be accomplished with a pair of one-sided hypothesis tests:

Test 1: H0: $\delta < = -\Delta$ versus H1: $\delta > -\Delta$

Test 2: H0: $\delta > = \Delta$ versus H1: $\delta < \Delta$

If we reject both of these tests, then we establish that $-\Delta < \delta < \Delta$, or equivalently that $|\delta| < \Delta$.

Unequal Variances

If the variances of the two populations are not the same, neither the t-test nor the permutation test will yield exact significance levels despite pronouncements to the contrary of numerous experts regarding the permutation tests.

Rule 1: If the underlying distribution is known, make use of it.

Some older textbooks recommend the use of an arcsine transformation when the data are drawn from a binomial distribution and a square root transformation when the data are drawn from a Poisson distribution. The resultant p-values are still only approximations and, in any event, lead to suboptimal tests.

The optimal test for comparing two binomial distributions is Fisher's Exact Test, and the optimal test for comparing two Poisson distributions is based on the binomial distribution [see, e.g., Lehmann, 1986, Chapter 5, Section 5].

> Rule 2: More important than comparing mean values can be determining why the variances of the populations are different.

There are numerous possible solutions for the Behrens–Fisher problem of unequal variances in the treatment groups. These include the following:

- Wilcoxon test; the use of the ranks in the combined sample reduces the impact (though not the entire effect) of the difference in variability between the two samples.
- Generalized Wilcoxon test; see O'Brien [1988].
- Procedure described in Manly and Francis [1999].
- Procedure described in Chapter 7 of Weerahandi [1995].
- Procedure described in Chapter 10 of Pesarin [2001].
- Bootstrap. See the following section on dependent observations.
- Permutation test. Phillip Good conducted simulations for sample sizes between 6 and 12 drawn from normally distributed populations. The populations in these simulations had variances that differed by up to a factor of 5, and nominal p-values of 5% were accurate to within 1.5%.

Hilton [1996] compared the power of the Wilcoxon test, the O'Brien test, and the Smirnov test in the presence of both location shift and scale (variance) alternatives. As the relative influence of the difference in variances grows, the O'Brien test is most powerful. The Wilcoxon test loses power in the face of different variances. If the variance ratio is $4:1$, the Wilcoxon test is not trustworthy.

One point is unequivocal. William Anderson writes:

> The first issue is to understand *why* the variances are so different, and what does this mean to the patient. It may well be the case that a new treatment is not appropriate because of higher variance, even if the difference in means is favorable. This issue is important whether the difference was anticipated. Even if the regulatory agency does not raise the issue, I want to do so internally.

David Salsburg agrees.

> If patients have been assigned at random to the various treatment groups, the existence of a significant difference in any parameter of the distribution suggests that there is a difference in treatment effect. The problem is not how to compare the means but how to determine what aspect of this difference is relevant to the purpose of the study.

Since the variances are significantly different, I can think of two situations where this might occur:

1. In many measurements there are minimum and maximum values that are possible, for example, the Hamilton Depression Scale, or the number of painful joints in arthritis. If one of the treatments is very effective, it will tend to push values into one of the extremes. This will produce a change in distribution from a relatively symmetric one to a skewed one, with a corresponding change in variance.

2. The experimental subjects may represent a mixture of populations. The difference in variance may occur because the effective treatment is effective for only a subset of the population. A locally most powerful test is given in Conover and Salsburg [1988].

Dependent Observations

The preceding statistical methods are not applicable if the observations are interdependent. There are five cases in which, with some effort, analysis may still be possible: repeated measures, clusters, known or equal pairwise dependence, a moving average or autoregressive process,[8] and group-randomized trials.

Repeated Measures. Repeated measures on a single subject can be dealt with in a variety of ways, including treating them as a single multivariate observation. Good [2001, Section 5.6] and Pesarin [2001, Chapter 11] review a variety of permutation tests for use when there are repeated measures.

Another alternative is to use one of the standard modeling approaches such as random- or mixed-effects models or generalized estimating equations (GEEs). See Chapter 13 for a full discussion.

Clusters. Occasionally, data will have been gathered in clusters from families and other groups who share common values, work, or leisure habits. If stratification is not appropriate, treat each cluster as if it were a single observation, replacing individual values with a summary statistic such as an arithmetic average [Mosteller and Tukey, 1977].

Cluster-by-cluster means are unlikely to be identically distributed, having variances, for example, that will depend on the number of individuals that make up the cluster. A permutation test based on these means would not be exact.

If there is a sufficiently large number of such clusters in each treatment group, the *bootstrap*, defined in Chapters 3 and 7, is the appropriate method of analysis. In this application, bootstrap samples are drawn on the clusters rather than the individual observations.

With the bootstrap, the sample acts as a surrogate for the population. Each time we draw a pair of bootstrap samples from the original sample, we compute the difference in means. After drawing a succession of such samples, we'll have some idea of what the distribution of the difference in means would be were we to take repeated pairs of samples from the population itself.

---

[8]For a discussion of these cases, see Brockwell and Davis [1987].

As a general rule, resampling should reflect the null hypothesis, according to Young [1986] and Hall and Wilson [1991]. Thus, in contrast to the bootstrap procedure used in estimation (see Chapters 3 and 7), each pair of bootstrap samples should be drawn from the *combined sample* taken from the two treatment groups. Under the null hypothesis, this will not affect the results; under an alternative hypothesis, the two bootstrap sample means will be closer together than they would if drawn separately from the two populations. The difference in means between the two samples that were drawn originally should stand out as an extreme value.

Hall and Wilson [1991] also recommend that the bootstrap be applied only to statistics that, for very large samples, will have distributions that do not depend on any unknowns.[9] In the present example, Hall and Wilson recommend the use of the *t*-statistic, rather than the simple difference of means, as leading to a test that is both closer to exact and more powerful.

Suppose we draw several hundred such bootstrap samples with replacement from the combined sample and compute the *t*-statistic each time. We would then compare the original value of the test statistic, Student's *t* in this example, with the resulting bootstrap distribution to determine what decision to make.

Pairwise Dependence. If the covariances are the same for each pair of observations, then the permutation test described previously is an exact test if the observations are normally distributed [Lehmann, 1986] and is almost exact otherwise.

Even if the covariances are not equal, if the covariance matrix is nonsingular, we may use the inverse of this covariance matrix to transform the original (dependent) variables to independent (and hence exchangeable) variables. After this transformation, the assumptions are satisfied so that a permutation test can be applied. This result holds even if the variables are collinear. Let R denote the rank of the covariance matrix in the singular case. Then there exists a projection onto an R-dimensional subspace where R normal random variables are independent. So, if we have an N-dimensional $(N > R)$ correlated and singular multivariate normal distribution, there exists a set of R linear combinations of the original N variables so that the R linear combinations are each univariate normal and independent.

The preceding is only of theoretical interest unless we have some independent source from which to obtain an estimate of the covariance matrix. If we use the data at hand to estimate the covariances, the estimates will be interdependent and so will the transformed observations.

Moving Average or Autoregressive Process. These cases are best treated by the same methods and are subject to the caveats described in Part III of this book.

Group Randomized Trials.[10] Group randomized trials (GRTs) in public health research typically use a small number of randomized groups with a relatively large number of participants per group. Typically, some naturally occurring groups are

---

[9]Such statistics are termed asymptotically pivotal.

[10]This section has been abstracted with permission from *Annual Reviews* from Feng et al. [2001], from which all quotes in this section are taken.

targeted: work sites, schools, clinics, neighborhoods, and even entire towns or states. A group can be assigned to either the intervention or control arm but not both; thus, the group is nested within the treatment. This contrasts with the approach used in multi-center clinical trials, in which individuals within groups (treatment centers) may be assigned to any treatment.

GRTs are characterized by a positive correlation of outcomes within a group and by the small number of groups.

There is positive intraclass correlation (ICC), between the individuals' target-behavior outcomes within the same group. This can be due in part to the differences in character-istics between groups, to the interaction between individuals within the same group, or (in the presence of interventions) to commonalities of the intervention experienced by an entire group. Although the size of the ICC in GRTs is usually very small (e.g., in the Working Well Trial, between 0.01–0.03 for the four outcome variables at baseline), its impact on the design and analysis of GRTs is substantial.

The sampling variance for the average responses in a group is $(\sigma^2/n) * [1 + (n - 1)\sigma)]$, and that for the treatment average with k groups and n individuals per group is $(\sigma^2/n) * [1 + (n - 1)\sigma]$, not the traditional σ^2/n and $\sigma^2/(nk)$, respectively, for uncorre-lated data.

The factor $1 + (n - 1)\sigma$ is called the variance inflation factor (VIF), or design effect. Although σ in GRTs is usually quite small, the VIFs could still be quite large because VIF is a function of the product of the correlation and group size n.

For example, in the Working Well Trial, with $\sigma = 0.03$ for [the] daily number of fruit and vegetable servings, and an average of 250 workers per work site, VIF = 8.5. In the pre-sence of this deceivingly small ICC, an 8.5-fold increase in the number of participants is required to maintain the same statistical power as if there were no positive correlation. Ignoring the VIF in the analysis would lead to incorrect results: variance estimates for group averages that are too small. See Table 6.5.

To be appropriate, an analysis method of GRTs must acknowledge both the ICC and the relatively small number of groups. Three primary approaches are used:

1. Generalized linear mixed models (GLMM). This approach, implemented in SAS Macro GLIMMIX and SAS PROC MIXED, relies on an assumption of normality.

2. GEE. Again, this approach assumes asymptotic normality for conducting infer-ence, a good approximation only when the number of groups is large.

3. Randomization-based inference. Unequal-size groups will result in unequal var-iances of treatment means, resulting in misleading p-values. To be fair, "Gail et al. [1996] demonstrate that in GRTs, the permutation test remains valid (exact or near exact in nominal levels) under almost all practical situations, including unbalanced group sizes, as long as the number of groups are equal between treatment arms or equal within each block if blocking is used."

The drawbacks of all three methods, including randomization-based inference if corrections are made for covariates, are the same as those for other methods of regression, as detailed in Chapters 8 and 9.

TABLE 6.5. Comparison of Different Analysis Methods for Inference on Treatment Effect $\hat{\beta}^a$

Method 10^2	$\hat{\beta}$ (10^2 SE)	p-Value	$\hat{p}$
Fruit/vegetable			
GLIM (independent)	−6.9 (2.0)	0.0006	
GEE (exchangeable)	−6.8 (2.4)	0.0052	0.0048
GLMM (random intercept) df D 12^b	−6.7 (2.6)	0.023	0.0077
Permutation	−6.1 (3.4)	0.095	
t-Test (group-level)	−6.1 (3.4)	0.098	
Permutation (residual)	−6.3 (2.9)	0.052	
Smoking			
GLIM (independent)	−7.8 (12)	0.53	
GEE (exchangeable)	−6.2 (20)	0.76	0.0185
GLMM (random intercept) df D 12^b	−13 (21)	0.55	0.020
Permutation	−12 (27)	0.66	
t-Test (group level)	−12 (27)	0.66	
Permutation (residual)	−13 (20)	0.53	

[a]Using Seattle 5-a-day data with 26 work sites ($K=13$) and an average of 87 (n_i ranges from 47 to 105) participants per work site. The dependent variables are ln (daily servings of fruit and vegetable C_1) and smoking status. The study design is matched pair, with two cross-sectional surveys at baseline and 2-year follow-up. Pairs identification, work sites nested within treatment, intervention indicator, and baseline work-site mean fruit-and-vegetable intake are included in the model. Pairs and work sites are random effects in GLMM (generalized linear mixed models). We used SAS PROC GENMOD for GLIM (linear regression and generalized linear models) and GEE (generalized estimating equations) (logistic model for smoking data) and SAS PROCMIXED (for fruit/vegetable data) or GLMMIX (logistic regression for smoking data) for GLMM; permutation tests (logit for smoking data) were programmed in SAS.

[b]Degrees of freedom (df) = 2245 in SAS output if work site is not defined as being nested within treatment.

Source: Reprinted with permission from *Annual Reviews* from Feng et al. [2001].

Nonsystematic Dependence. If the observations are interdependent and fall into none of the preceding categories, then the experiment is fatally flawed. Your efforts would be best expended on the design of a cleaner experiment. Or, as J.W. Tukey remarked on more than one occasion, "If a thing is not worth doing, it is not worth doing well."

Don't Let Your Software Do Your Thinking

1. Most statistical software comes with built-in defaults, such as a two-sided test at the 5% significance level. If altered, subsequent uses will default back to the previously used specifications. But what if these settings are not appropriate for your particular application? We know of one statistician who advised his company to take twice as many samples as necessary (at twice the investment in money and time) simply because he'd allowed the software to make the settings. Always verify that the current default settings of your statistical software are appropriate before undertaking an analysis or a sample-size determination.

2. It is up to you, not your software, to verify that all the necessary assumptions are satisfied. The fact that your software yields a *p*-value does not mean that you performed the appropriate analysis.

COMPARING VARIANCES

Testing for the equality of the variances of two populations is a classic problem with many not-quite-exact, not-quite-robust, not-quite-powerful-enough solutions. Sukhatme [1958] lists 4 alternative approaches and adds a fifth of his own; Miller [1968] lists 10 alternatives and compares 4 of them with a new test of his own; Conover et al. [1981] list and compare 56 tests; and Balakrishnan and Ma [1990] list and compare 9 tests with one of their own.

None of these tests proves satisfactory in all circumstances, for each requires that two or more of the following four conditions be satisfied:

1. The observations are normally distributed.
2. The location parameters of the two distributions are the same or differ by a known quantity.
3. The two samples are equal in size.
4. The samples are large enough that asymptotic approximations to the distribution of the test statistic are valid.

As an example, the first published solution to this classic testing problem is the *z*-test proposed by Welch [1937] based on the ratio of the two sample variances. If the observations are normally distributed, this ratio has the *F*-distribution, and the test whose critical values are determined by the *F*-distribution is uniformly most powerful among all unbiased tests [Lehmann, 1986, Section 5.3]. But with even small deviations from normality, significance levels based on the *F*-distribution are grossly in error [Lehmann, 1986, Section 5.4].

Box and Anderson [1955] propose a correction to the *F*-distribution for "almost" normal data based on an asymptotic approximation to the permutation distribution of the *F*-ratio. Not surprisingly, their approximation is close to correct only for normally distributed data or for very large samples. The Box–Anderson statistic results in an error rate of 21%, twice the desired value of 10%, when two samples of size 15 are drawn from a gamma distribution with four degrees of freedom.

A more recent permutation test [Bailor, 1989] based on complete enumeration of the permutation distribution of the sample *F*-ratio is exact only when the location parameters of the two distributions are known or are known to be equal.

The test proposed by Miller [1968] yields conservative Type I errors, less than or equal to the declared error, unless the sample sizes are unequal. A 10% test with samples of size 12 and 8 taken from normal populations yielded Type I errors 14% of the time.

Fligner and Killeen [1976] propose a permutation test based on the sum of the absolute deviations from the combined sample mean. Their test may be appropriate when the medians of the two populations are equal but can be virtually worthless otherwise, accepting the null hypothesis up to 100% of the time. In the first edition, Good [2001] proposed a test based on permutations of the absolute deviations from the individual sample medians; this test, alas, is only asymptotically exact and even then only for approximately equal sample sizes, as shown by Baker [1995].

To compute the primitive bootstrap introduced by Efron [1979], we would take successive pairs of samples—one of n observations from the sampling distribution F_n which assigns mass $1/n$ to the values $\{X_i: i = 1, \ldots n\}$, and one of m observations from the sampling distribution G_m which assigns mass $1/m$ to the values $\{X_j: j = n+1, \ldots, n+m\}$, and compute the ratio of the sample variances:

$$R = \frac{s_n^2/(n-1)}{s_m^2/(m-1)}$$

We would use the resultant bootstrap distribution to test the hypothesis that the variance of F equals the variance of G against the alternative that the variance of G is larger. Under this test, we reject the null hypothesis if the $100(1-\alpha)$ percentile is less than 1.

This primitive bootstrap and the associated confidence intervals are close to exact only for very large samples with hundreds of observations. More often, the true coverage probability is larger than the desired value.

Two corrections yield vastly improved results. First, for unequal-size samples, Efron [1982] suggests that more accurate confidence intervals can be obtained using the test statistic

$$R' = \frac{s_n^2/n}{s_m^2/m}$$

Second, applying the bias and acceleration corrections described in Chapter 3 to the bootstrap distribution of R' yields almost exact intervals.

Lest we keep you in suspense, a distribution-free exact and more powerful test for comparing variances can be derived based on the permutation distribution of Aly's statistic.

This statistic, proposed by Aly [1990], is

$$\delta = \sum_{i=1}^{m-1} i(m-i)(X_{(i+1)} - X_{(i)})$$

where $X_{(1)} \leq X_{(2)} \leq \cdots \leq X_{(m)}$ are the order statistics of the first sample.

Suppose we have two sets of measurements, 121, 123, 126, 128.5, 129 and, in a second sample, 153, 154, 155, 156, 158. We replace these with the deviations $z_{1i} = X_{(i+1)} - X_{(i)}$ or 2, 3, 2.5, .5 for the first sample and $z_{2i} = 1, 1, 1, 2$ for the second.

The original value of the test statistic is $8 + 18 + 15 + 2 = 43$. Under the hypothesis of equal dispersions in the two populations, we can exchange labels between z_{1i} and z_{2i} for any or all of the values of i. One possible rearrangement of the labels on the deviations puts $\{2, 1, 1, 2\}$ in the first sample, which yields a value of $8 + 6 + 6 + 8 = 28$.

There are $2^4 = 16$ rearrangements of the labels in all, of which only one $\{2, 3, 2.5, 2\}$ yields a larger value of Aly's statistic than the original observations. A one-sided test would have 2 out of 16 rearrangements as extreme as or more extreme than the original; a two-sided test would have 4. In either case, we would accept the null hypothesis, though the wiser course would be to defer judgment until we have taken more observations.

If our second sample is larger than the first, we have to resample in two stages. First, we select a subset of m values at random without replacement from the n observations in the second, larger sample and compute the order statistics and their differences. Last, we examine all possible values of Aly's measure of dispersion for permutations of the combined sample, as we did when the two samples were equal in size, and compare Aly's measure for the original observations with this distribution. We repeat this procedure several times to check for consistency.

Good [2005] proposed a permutation test based on the sum of the absolute values of the deviations about the median. First, we compute the median for each sample; next, we replace each of the remaining observations by the square of its deviation about its sample median; last, in contrast to the test proposed by Brown and Forsythe [1974], we discard the redundant linearly dependent value from each sample.

Suppose the first sample contains the observations $x_{11}, \ldots, x_{1n_1}$ whose median is M_1; we begin by forming the deviates $\{x'_{1j} = |x_{1j} - M_1|\}$ for $j = 1, \ldots, n_1$. Similarly, we form the set of deviates $\{x'_{2j}\}$ using the observations in the second sample and their median.

If there is an odd number of observations in the sample, then one of these deviates must be zero. We can't get any information out of a zero, so we throw it away. In the event of ties, should there be more than one zero, we still throw only one away. If there is an even number of observations in the sample, then two of these deviates (the two smallest ones) must be equal. We can't get any information out of the second one that we didn't already get from the first, so we throw it away.

Our new test statistic, S_G, is the sum of the remaining $n_1 - 1$ deviations in the first sample, that is, $S_G = \sum_{j=1}^{n_1-1} x'_{1j}$.

We obtain the permutation distribution for S_G and the cutoff point for the test by considering all possible rearrangements of the remaining deviations between the first and second samples.

To illustrate the application of this method, suppose the first sample consists of the measurements 121, 123, 126, 128.5, 129.1 and the second sample consists of the measurements 153, 154, 155, 156, 158. Thus, after eliminating the zero value, $x'_{11} = 5$, $x'_{12} = 3$, $x'_{13} = 2.5$, $x'_{14} = 3.1$, and $S_G = 13.6$. For the second sample, $x'_{21} = 2$, $x'_{22} = 1$, $x'_{23} = 1$, and $x'_{24} = 3$.

In all, there are $\binom{8}{4} = 70$ arrangements, of which only three yield values of the test statistic as extreme as or more extreme than our original value. Thus, our p-value is $3/70 = 0.043$, and we conclude that the difference between the dispersions of the two manufacturing processes is statistically significant at the 5% level.

As there is still a weak dependency among the remaining deviates within each sample, they are only asymptotically exchangeable. Tests based on S_G are alternately conservative and liberal, according to Baker [1995], in part because of the discrete nature of the permutation distribution unless

1. The ratio of the sample sizes n, m is close to 1.
2. The only other difference between the two populations from which the samples are drawn is that they might have different means, that is, $F_2[x] = F_1[(x - \delta)/\sigma]$.

The preceding test is easily generalized to the case of K samples from K populations. Such a test would be of value as a test for homoskedacity as a preliminary to a k-sample analysis for a difference in means among test groups.

First, we create K sets of deviations about the sample medians and make use of the test statistic

$$S = \sum_{k=1}^{K} \left(\sum_{j=1}^{n_1-1} x'_{1j} \right)^2$$

The choice of the square of the inner sum ensures that this statistic takes its largest value when the largest deviations are all together in one sample after relabeling.

To generate the permutation distribution of S, we again have two choices. We may consider all possible rearrangements of the sample labels over the K sets of deviations. Or, if the samples are equal in size, we may first order the deviations within each sample, group them according to rank, and then rearrange the labels within each ranking.

Again, this latter method is directly applicable only if the K samples are equal in size, and, again, this is unlikely to occur in practice. We will have to determine a confidence interval for the p-value for the second method via a bootstrap in which we first select samples from samples (without replacement) so that all samples are equal in size. While we wouldn't recommend doing this test by hand, once programmed, it still takes less than a second on last year's desktop.

MATCH SIGNIFICANCE LEVELS BEFORE PERFORMING POWER COMPARISONS

When we studied the small-sample properties of parametric tests based on asymptotic approximations that had performed well in previously published power comparisons, we uncovered another major error in statistics: the failure to match significance levels before performing power comparisons. Asymptotic approximations to cutoff values were used rather than exact values or near estimates.

When a statistical test takes the form of an interval, that is, if we reject when $S < c$ and accept otherwise, then power is a nondecreasing function of significance level; a test based on an interval may have greater power at the 10% significance level than a second different test evaluated at the 5% significance level, even though the second test is uniformly more powerful than the first. To see this, let H denote the primary hypothesis and K an alternative hypothesis.

If $\Pr\{S < c \mid H\} = \alpha < \alpha' = \Pr\{S < c' \mid H\}$, then $c < c'$, and $\beta = \Pr\{S < c \mid K\} \leq \Pr\{S < c' \mid K\} = \beta'$.

Consider a second statistical test depending on S via the monotone increasing function h, where we reject if $h[S] < d$. If the cutoff values $d < d'$ correspond to the same significance levels $\alpha < \alpha'$, then $\beta < \Pr\{h[S] < d \mid K\} < \beta'$. Even though the second test is more powerful than the first at level α, this will not be apparent if we substitute an approximate cutoff point c' for an exact one c when comparing the two tests.

To ensure matched significance levels in your own power comparisons, proceed in two stages: First, use simulations to derive exact cutoff values. Then use these derived cutoff values in determining power. Using this approach, we were able to show that an exact permutation test based on Aly's statistic was more powerful for comparing variances than any of the numerous published inexact parametric tests.

Normality is a myth; there never has, and never will be a normal distribution.

—Geary [1947, p. 241]

COMPARING THE MEANS OF k SAMPLES

Although, the traditional one-way analysis of variance based on the F-ratio

$$\frac{\sum_{i=1}^{I} n_i (X_{i.} - X_{..})^2 / (I - 1)}{\sum_{i=1}^{I} \sum_{j=1}^{n_i} (X_{ij} - X_{i.})^2 / (N - I)}$$

is highly robust, it has four major limitations:

1. Its significance level is dependent on the assumption of normality. Problems occur when data are drawn from distributions that are highly skewed or heavy in the tails. Still, the F-ratio test is remarkably robust to minor deviations from normality.

2. Not surprisingly, lack of normality also affects the power of the test, rendering it suboptimal.

3. The F-ratio is optimal for losses that are proportional to the square of the error and is suboptimal otherwise.

4. The F-ratio is an omnibus statistic offering all-around power against many alternatives but no particular advantage against any one of them. For example, it is suboptimal for testing against an ordered dose response when a test based on the correlation would be preferable.

A permutation test is preferred for the k-sample analysis [Good and Lunneborg, 2005]. These tests are distribution free (though the variances must be the same for all treatments). They are at least as powerful as the analysis of variance. And you can choose the test statistic that is optimal for a given alternative and loss function and not be limited by the availability of tables.

We take as our model $X_{ij} = \alpha_i + \varepsilon_{jj}$, where $i = 1, \ldots, I$ denotes the treatment, and $j = 1, \ldots, n_i$. We assume that the error terms $\{\varepsilon_{jj}\}$ are i.i.d.

We consider two loss functions. In one of them, the losses associated with overlooking a real treatment effect, a Type II error, are proportional to the sum of the squares of the treatment effects α_i^2 (LS). In the other, the losses are proportional to the sum of the absolute values of the treatment effects, $|\alpha_i|$ (LAD).

Our hypothesis, a null hypothesis, is that the differential treatment effects, the $\{\alpha_i\}$, are all zero. We will also consider two alternative hypotheses: K_U, that at least one of the differential treatment effects α_i is not zero, and K_O, that exactly one of the differential treatment effects α_i is not zero.

For testing against K_U with the LS loss function, Good [2002, p. 126] recommends the use of the statistic $F_2 = \sum_i (\sum_j X_{ij})^2$, which is equivalent to the F-ratio once terms that are invariant under permutations are eliminated.

We compared the parametric and permutation versions of this test when the data were drawn from a mixture of normal distributions. The difference between the two in power is increased when the design is unbalanced. For example, the following experiment was simulated 4000 times:

- A sample of size three was taken from a mixture of 70%$N(0, 1)$ and 30%$N(1, 1)$.
- A sample of size four was taken from a mixture of 70%$N(0.5, 1)$ and 30%$N(1.5, 1.5)$.
- A sample of size five was taken from a mixture of 70%$N(1, 1)$ and 30%$N(2, 2)$.

Note that such mixtures are extremely common in experimental work. The parametric test in which the F-ratio is compared with an F-distribution had a power of 18%. The permutation test in which the F-ratio is compared with a permutation distribution had a power of 31%.

For testing against K_U with the LAD loss function, Good [2002, p. 126] recommends the use of the statistic $F_1 = \sum_i |\sum_j X_{ij}|$.

For testing against K_O, first denote by $\overline{X}_i$ the mean of the ith sample and by $\overline{X}^i$ the mean of all observations excluding those in the ith sample. A possible test statistic would be the maximum of the differences $|\overline{X}^i - \overline{X}_i|$.

A permutation test based on the original observations is appropriate only if one can assume that under the null hypothesis the observations are identically distributed in each of the populations from which the samples are drawn. If we cannot make this assumption, we will need to transform the observations, throwing away some of the information about them so that the distributions of the transformed observations are identical.

For example, for testing against K_O, Lehmann [1998, p. 372] recommends the use of the Jonckheere–Terpstra statistic, the number of pairs in which an observation from one group is less than an observation from a higher-dose group. The penalty we pay for using this statistic and ignoring the actual values of the observations is a marked reduction in power for small samples and a less pronounced loss for larger ones.

If there are just two samples, the test based on the Jonckheere–Terpstra statistic is identical to the Mann–Whitney test. For very large samples with identically distributed observations in both samples, 100 observations would be needed with this test to obtain the same power as a permutation test based on the original values of 95 observations. This is not a price one would want to pay in human or animal experiments.

SUBJECTIVE DATA

Student's t and the analysis of variance are based on mathematics that requires the dependent variable to be measured on an interval or ratio scale so that its values can be meaningfully added and subtracted. But what does it mean if one subtracts the subjective data value "Indifferent" from the subjective data value "Highly preferable"? The mere fact that we have entered the data into the computer on a Likert scale, such as a 1 for "Highly preferable" and a 3 for "Indifferent" does not actually endow our preferences with those relative numeric values.

Unfortunately, the computer thinks it does, and if asked to compute a mean preference, it will add the numbers it has stored and divide by the sample size. It will even compute a t statistic and a p-value if these are requested. But this does not mean that either is meaningful.

Of course, you are welcome to ascribe numeric values to subjective data, provided that you spell out exactly what you have done and realize that the values you ascribe may be quite different from the ones that I or some other investigator might attribute to precisely the same data.

INDEPENDENCE VERSUS CORRELATION

Recent simulations reveal that the classic test based on Pearson correlation is distribution free. Still, too often we treat a test of the correlation between two variables X and Y as if it were a test of their independence. X and Y can have a zero correlation coefficient, yet be totally dependent (e.g., $y = x^2$).

For example, even when the expected value of Y is independent of the expected value of X, the variance of Y might be directly proportional to the variance of X. Of course, if we'd plotted the data, we'd have spotted this right away.

Many variables exhibit circadian rhythms. Yet, the correlation of such a variable with time when measured over the course of 24 hours would be zero. This is because correlation really means "linear correlation," and the behavior of the diurnal rhythm is far from linear. Of course, this too would have been obvious had we drawn a graph rather than let the computer do the thinking for us.

Yet another, not uncommon, example would be when X is responsible for the size of a change in Y but a third variable, not part of the study, determines the direction of the change.

HIGHER-ORDER EXPERIMENTAL DESIGNS

The two principal weaknesses of the analysis of variance are as follows:

1. The various tests of significance are *not* independent of one another, as they are based on statistics that share a common denominator.
2. Undefined confounding variables may create the illusion of a relationship or may mask an existing one.

When we randomly assign subjects (or plots) to treatment, we may inadvertently assign all males, say, to one of the treatments. The result might be the illusion of a treatment effect that really arises from a sex effect. For example, the following table suggests that there exists a statistically significant difference between treatments.

Source of Variation	Sum of Squares	DF	Mean Square	F	p-Value
Between groups	29,234.2	3	9744.73	3.43	0.038
Within groups	53,953.6	19	2839.66		
Corrected total	83,187.6	22			

But suppose we were to analyze the same data correcting for sex and obtain the following, in which we no longer observe a statistically significant difference between treatment groups.

Source of Variation	Sum of Squares	DF	Mean Square	F	p-Value
Treatment	24,102.2	3	8034.07	2.84	0.067
Sex	8200.5	1	8200.5	2.90	0.106
Within groups	50,884.9	18	2826.94		
Corrected total	83,187.6	22			

Errors in Interpretation

As noted previously, one of the most common statistical errors is to assume that because an effect is not statistically significant, it does not exist. One of the most common errors in using analysis of variance is to assume that because a factor such as sex does not yield a significant p-value, we may eliminate it from the model. Had we done so in the above example, we would have observed a statistically significant difference among treatments that was actually due to the unequal distribution of the sexes among the various treatments.

The process of eliminating nonsignificant factors one by one from an analysis of variance means that we are performing a series of tests rather than a single test; thus, the actual significance level is larger than the declared significance level.

Multifactor Designs

Further problems arise when one comes to interpret the output of three-way, four-way, and higher-order designs. Suppose a second- or higher-order interaction is statistically significant; how is this to be given a practical interpretation? Some authors suggest that one write, "Factor C moderates the effect of Factor A on Factor B," as if this sentence actually had discernible meaning. Among the obvious alternative interpretations of a statistically significant higher-order interaction are the following:

- An example of a Type I error.
- A defect in the formulation of the additive model; perhaps one ought to have employed $f(X)$ in place of X or $g(X, Y)$ in place of $X * Y$.

Still, it is clear that there are situations in which higher-order interactions have real meaning. For example, plants require nitrogen, phosphorus, and potassium in sufficient concentrations to grow. Remove any one component and the others will prove inadequate to sustain growth—a clear-cut example of a higher-order interaction.

To avoid ambiguities, one must either treat multifactor experiments purely as pilot efforts and guides to further experimentation or undertake such experiments only after one has gained a thorough understanding of interactions via one- and two-factor experiments. See the discussion in Chapter 11 on building a successful model.

On the plus side, the parametric analysis of variance is remarkably robust with respect to data from nonnormal distributions [Jagers, 1980]. As with the k-sample comparison, it should be remembered that the tests for main effects in the analysis of variance are omnibus statistics offering all-around power against many alternatives but no particular advantage against any one of them.

Judicious use of contrasts can provide more powerful tests. For example, one can obtain a one-sided test of the row effect in a 2xCx ... design by testing the contrast $\overline{X}_{1...} - \overline{X}_{2...}$ or a test of an ordered row effect in an RxCx ... design by testing the contrast $\sum_j a_j X_j ...$, where $\sum a_j = 0$ and the a_j are increasing in j. Note: These contrasts must be specified before examining the data. Otherwise, there will be a loss of power due to the need to correct for multiple tests.

Two additional caveats apply to the parametric analysis of variance approach to the analysis of two-factor experimental design:

1. The sample sizes must be the same in each cell; that is, the design must be balanced.
2. A test for interaction must precede any test for main effects.

Alas, these same caveats apply to the permutation tests. Let us see why.

Imbalance in the design will result in the confounding of main effects with interactions. Consider the following two-factor model for crop yield:

$$X_{ijk} = \mu + \alpha_i + \beta_j + \gamma_{ij} + \varepsilon_{jjk}$$

Now suppose that the observations in a two-factor experimental design are normally distributed, as in the following diagram taken from Cornfield and Tukey [1956]:

$$\frac{N(0, 1) \,|\, N(2, 1)}{N(2, 1) \,|\, N(0, 1)}$$

There are no main effects in this example. Both row means and both column means have the same expectations, but there is a clear interaction represented by the two non-zero off-diagonal elements.

If the design is balanced, with equal numbers per cell, the lack of significant main effects and the presence of a significant interaction should and will be confirmed by our analysis. But suppose that the design is not in balance—that for every 10 observations in the first column, we have only 1 observation in the second. Because of this imbalance, when we use the F-ratio or an equivalent statistic to test for the main effect, we will uncover a false "row" effect that is actually due to the interaction between rows and columns. The main effect is *confounded* with the interaction.

If a design is unbalanced, as in the preceding example, we cannot test for a "pure" main effect or a "pure" interaction. But we may be able to test for the combination of a main effect with an interaction by using the statistic that we would use to test for the main effect alone. This combined effect will not be confounded with the main effects of other unrelated factors.

Whether or not the design is balanced, the presence of an interaction may zero out a cofactor-specific main effect or make such an effect impossible to detect. More important, the presence of a significant interaction may render the concept of a single main effect meaningless. For example, suppose we decide to test the effect of fertilizer and sunlight on plant growth. With too little sunlight, a fertilizer would be completely ineffective. Its effects appear only when sufficient sunlight is present. Aspirin and warfarin can both reduce the likelihood of repeated heart attacks when used alone; you don't want to mix them!

Gunter Hartel offers the following example: Using five observations per cell and random normals, as indicated in Cornfield and Tukey's [1956] diagram, a two-way analysis of variance without interaction yields the following results:

Source	DF	Sum of Squares	F-Ratio	Prob $> F$
Row	1	0.15590273	0.0594	0.8104
Col	1	0.10862944	0.0414	0.8412
Error	17	44.639303		

Adding the interaction term yields

Source	DF	Sum of Squares	F-Ratio	Prob $> F$
Row	1	0.155903	0.1012	0.7545
Col	1	0.108629	0.0705	0.7940
Row * col	1	19.986020	12.9709	0.0024
Error	16	24.653283		

Expanding the first row of the experiment to have 80 observations rather than 10, the main-effects-only table becomes

Source	DF	Sum of Squares	F-Ratio	Prob > F
Row	1	0.080246	0.0510	0.8218
Col	1	57.028458	36.2522	<.0001
Error	88	138.43327		

But with the interaction term, it is

Source	DF	Sum of Squares	F-Ratio	Prob > F
Row	1	0.075881	0.0627	0.8029
Col	1	0.053909	0.0445	0.8333
Row * col	1	33.145790	27.3887	<.0001
Error	87	105.28747		

The standard permutation tests for main effects and interactions in a multifactor experimental design are also correlated, as the residuals (after subtracting main effects) are not exchangeable even if the design is balanced [Lehmann and D'Abrera, 1988]. To see this, suppose our model is $X_{ijk} = \mu + \alpha_i + \beta_j + \gamma_{ij} + \varepsilon_{ijk}$, where $\sum \alpha_i = \sum \beta_j = \sum_i \gamma_{ij} = \sum_j \gamma_{ij} = 0$.

Eliminating the main effects in the traditional manner, that is, setting $X'_{ijk} = X_{ijk} - \overline{X}_{i..} - \overline{X}_{.j.} + \overline{X}...$, one obtains the test statistic

$$I = \sum_i \sum_j \left(\sum_k X'_{ijk} \right)^2$$

first derived by Still and White [1981]. A permutation test based on the statistic I will not be exact. For even if the error terms $\{\varepsilon_{ijk}\}$ are exchangeable, the residuals $X'_{ijk} = \varepsilon_{ijk} - \overline{\varepsilon}_{i..} - \overline{\varepsilon}_{.j.} + \overline{\varepsilon}...$ are weakly correlated, the correlation depending on the subscripts.

The negative correlation between permutation test statistics works to their advantage only when just a single effect is present. Nonetheless, the literature is filled with references to permutation tests for the two-way and higher-order designs that produce misleading values. Included in this category are those permutation tests based on the ranks of the observations, such as the Kruskall–Wallace test that may be found in many statistics software packages.

Factorial Designs

Salmaso [2003] developed exact distribution-free tests for analyzing factorial designs.

Crossover Designs

Good and Xie [2008] developed an exact distribution-free test for analyzing crossover designs.

Unbalanced Designs

Unbalanced designs with unequal numbers per cell may result from unanticipated losses during the conduct of an experiment or survey (or from an extremely poor initial design). There are two approaches to their analysis.

First, if we have a large number of observations and only a small number are missing, we might consider imputing values to the missing observations, recognizing that the results may be somewhat tainted.

Second, we might bootstrap along one of the following lines:

- If only one or two observations are missing, create a balanced design by discarding observations at random; repeat to obtain a distribution of p-values [Baker, 1995].
- If there are actual holes in the design, so that there are missing combinations, create a test statistic that does not require the missing data. Obtain its distribution by bootstrap means. See Good [2005, pp. 138–140] for an example.

INFERIOR TESTS

Violation of assumptions can affect not only the significance level of a test but the power of the test as well; see Tukey and MacLaughlin [1963] and Box and Tiao [1964]. For example, while the significance level of the t-test is robust to departures from normality, the power of the t-test is not. Thus, the two-sample permutation test may always be preferable.

If blocking including matched pairs was used in the original design, then the same division into blocks should be employed in the analysis. Confounding factors such as sex, race, and diabetic condition can easily mask the effect we hoped to measure through the comparison of two samples. Similarly, an overall risk factor can be totally misleading [Gigerenzer, 2002]. Blocking reduces the differences between subjects so that differences between treatment groups stand out—if, that is, the appropriate analysis is used. Thus, paired data should always be analyzed with the paired t-test or its permutation equivalent, not with the group t-test.

To analyze a block design (e.g., where we have sampled separately from whites, blacks, and Hispanics), the permutation test statistic is $S = \sum_{b=1}^{B} \sum_{j} x_{bj}$, where x_{bj} is the jth observation in the control sample in the bth block, and the rearranging of labels between control and treated samples takes place separately and independently within each of the B blocks [Good, 2001, p. 124].

Blocking can also be used after the fact if you suspect the existence of confounding variables and if you measured the values of these variables as you were gathering data.[11]

Always be sure that your choice of statistic is optimal against the alternative hypotheses of interest for the appropriate loss function.

---

[11]This recommendation applies only to a test of efficacy for all groups (blocks) combined. p-Values for subgroup analyses performed after the fact are still suspect; see Chapter 1.

To avoid using an inferior, less sensitive, and possibly inaccurate statistical procedure, pay heed to another admonition from George Dyke [1997]: "The availability of 'user-friendly' statistical software has caused authors to become increasingly careless about the logic of interpreting their results, and to rely uncritically on computer output, often using the 'default option' when something a little different (usually, but not always, a little more complicated) is correct, or at least more appropriate."

MULTIPLE TESTS

When we perform multiple tests in a study, there may not be journal room (or interest) to report all the results, but we do need to report the total number of statistical tests performed so that readers can draw their own conclusions about the significance of the results that are reported.

We may also wish to correct the reported significance levels by using one of the standard correction methods for independent tests (e.g., the Bonferroni test, as described in Hsu [1996] and Aickin and Gensler [1996]; for resampling methods, see Westfall and Young [1993]).

Several statistical packages—SAS is a particular offender—print out the results of several dependent tests performed on the same set of data—for example, the t-test and the Wilcoxon test. We are not free to pick and choose. We must decide before we view the printout which test we will employ.

Let W_α denote the event that the Wilcoxon test rejects a hypothesis at the α significance level. Let P_α denote the event that a permutation test based on the original observations and applied to the same set of data rejects a hypothesis at the α significance level. Let T_α denote the event that a t-test applied to the same set of data rejects a hypothesis at the α significance level.

It is possible that W_α may be true when P_α and T_α are not, and so forth. As $\Pr\{W_\alpha$ or P_α or $T_\alpha \,|\, H\} \leq \Pr\{W_\alpha \,|\, H\} = \alpha$, we will have inflated the Type I error by choosing after the fact which test to report. Conversely, if our intent was to conceal a side effect by reporting that the results were not significant, we will inflate the Type II error and deflate the power β of our test by an after-the-fact choice as $\beta = \Pr\{\text{not } (W_\alpha \text{ and } P_\alpha \text{ and } T_\alpha) \,|\, K\} \leq \Pr\{W_\alpha \,|\, K\}$.

To repeat, we are not free to choose among tests; any such conduct is unethical.

Both the comparison and the test statistic must be specified in advance of examining the data.

BEFORE YOU DRAW CONCLUSIONS

Insignificance

If the p-value you observe is greater than your predetermined significance level, this may mean any or all of the following:

1. You've measured the wrong thing, measured it the wrong way, or used an inappropriate test statistic.

2. Your sample size was too small to detect an effect.

3. The effect you are trying to detect is not statistically significant.

Practical versus Statistical Significance

If the p-value you observe is less than your predetermined significance level, this does not necessarily mean that the effect you've detected is of practical significance—see, for example, the section on measuring equivalence. For this reason, as we discuss in Chapter 8, it is essential that you follow up any significant result by computing a confidence interval so that readers can judge for themselves whether the effect you've detected is of practical significance.

And do not forget that at the α-percent significance level, α-percent of your tests will be statistically significant by chance alone.

Missing Data

Before you draw conclusions, be sure that you have accounted for all missing data, interviewed nonresponders, and determined whether the data were missing at random or were specific to one or more subgroups.

During the Second World War, a group was studying airplanes returning from bombing Germany. They drew a rough diagram showing where the bullet holes were and recommended that those areas be reinforced. A statistician, Abraham Wald [1980],[12] pointed out that essential data were missing from the sample they were studying. What about the planes that didn't return from Germany?

When we think along these lines, we see that the two areas of the plane that had almost no bullet holes (where the wings and where the tail joined the fuselage) are crucial. Bullet holes in a plane are likely to be at random, occurring over the entire plane. Their absence in those two areas in returning bombers was diagnostic. Do the data missing from your experiments and surveys also have a story to tell?

Induction

> Behold! human beings living in an underground den, which has a mouth open towards the light and reaching all along the den; here they have been from their childhood, and have their legs and necks chained so that they cannot move, and can only see before them, being prevented by the chains from turning round their heads. Above and behind them a fire is blazing at a distance, and between the fire and the prisoners there is a raised way; and you will see, if you look, a low wall built along the way, like the screen which marionette players have in front of them, over which they show the puppets.

> And they see only their own shadows, or the shadows of one another, which the fire throws on the opposite wall of the cave.

> To them, I said, the truth would be literally nothing but the shadows of the images.
> —The Allegory of the Cave (Plato, *The Republic*, Book VII)

---

[12]This reference may be hard to obtain. Alternatively, see Mangel and Samaniego [1984].

Never assign probabilities to the true state of nature, but only to the validity of your own predictions.

A *p*-value does not tell us the probability that a hypothesis is true, nor does a significance level apply to any specific sample; the latter is a characteristic of our testing in the long run. Likewise, if all assumptions are satisfied, a confidence interval will in the long run contain the true value of the parameter a certain percentage of the time. But we cannot say with certainty in any specific case that the parameter does or does not belong to that interval [Neyman, 1961, 1977].

In our research efforts, the only statements we can make with God-like certainty are of the form "our conclusions fit the data." The true nature of the real world is unknowable. We can speculate, but never conclude.

The gap between the sample and the population will always require a leap of faith, for we understand only insofar as we are capable of understanding [Lonergan, 1992]. See also, the section on Deduction versus Induction in Chapter 2.

SUMMARY

Know your objectives in testing. Know your data's origins. Know the assumptions you feel comfortable with. Never assign probabilities to the true state of nature, but only to the validity of your own predictions. Collecting more and better data may be your best alternative.

TO LEARN MORE

For commentary on the use of wrong or inappropriate statistical methods, see Avram et al. [1985], Badrick and Flatman [1999], Berger et al. [2002], Bland and Altman [1995], Cherry [1998], Cox [1999], Dar, Serlin, and Omer [1997], Delucchi [1983], Elwood [1998], Felson, Cupples, and Meenan [1984], Fienberg [1990], Gore, Jones, and Rytter [1977], Lieberson [1985], MacArthur and Jackson [1977], McGuigan [1995], McKinney et al. [1989], Miller [1986], Padaki [1989], Welch and Gabbe [1996], Westgard and Hunt [1973], White [1979], and Yoccuz [1991].

Hunter and Schmidt [1997] emphasize why significance testing remains essential.

Guidelines for reviewers are provided by Altman [1998a], Bacchetti [2002], Finney [1997], Gardner, Machin, and Campbell [1986], George [1985], Goodman, Altman, and George [1998], International Committee of Medical Journal Editors [1997], Light and Pillemer [1984], Mulrow [1987], Murray [1988], Schor and Karten [1966], and Vaisrub [1985].

For additional comments on the effects of the violation of assumptions, see Box and Anderson [1955], Friedman [1937], Gastwirth and Rubin [1971], Glass, Peckham, and Sanders [1972], and Pettitt and Siskind [1981].

For the details of testing for equivalence, see Dixon [1998]. For a review of the appropriate corrections for multiple tests, see Tukey [1991].

For true tests of independence, see Romano [1989]. There are many tests for the various forms of dependence, such as quadrant dependence (Fisher's Exact Test), trend (correlation), and serial correlation [see, e.g., Maritz, 1996 and Manly, 1997].

For procedures with which to analyze factorial and other multifactor experimental designs, see Salmaso [2003] and Chapter 8 of Pesarin [2001].

Most of the problems with parametric tests reported here extend to and are compounded by multivariate analysis. For some solutions, see Chapter 9 of Good [2005], Chapter 6 of Pesarin [2001], and Pesarin [1990].

For a contrary view on adjustments of p-values in multiple comparisons, see Rothman [1990]. For a method for allocating Type I error among multiple hypotheses, see Moyé [2000].

Venn [1888] and Reichenbach [1949] are among those who've attempted to construct a mathematical bridge between what we observe and the reality that underlies our observations. To the contrary, extrapolation from the sample to the population is not a matter of applying Holmes-like deductive logic but entails a leap of faith. A careful reading of Locke [1700], Berkeley [1710], Hume [1748], and Lonergan [1992] is an essential prerequisite to the application of statistics.

For more on the contemporary view of induction, see Berger [2002] and Sterne, Smith, and Cox [2001]. The former notes that, "Dramatic illustration of the non-frequentist nature of p-values can be seen from the applet available at www.stat. duke.edu/~berger. The applet assumes one faces a series of situations involving normal data with unknown mean θ and known variance, and tests of the form H: $\theta = 0$ versus K: $\theta \neq 0$. The applet simulates a long series of such tests, and records how often H is true for p-values in given ranges."

7

MISCELLANEOUS STATISTICAL PROCEDURES

The greatest error associated with the use of statistical procedures is to make the assumption that a single statistical methodology can suffice for all applications.

From time to time, a new statistical procedure will be introduced or an old one revived along with the assertion that at last the definitive solution has been found. With a parallel to the establishment of new religions, at first the new methodology is reviled, even persecuted until, with growth in the number of its adherents, it can begin to attack and persecute the adherents of other, more established dogma in its turn.

During the preparation of this book, an editor of a statistics journal rejected an article of one of the authors on the sole grounds that it made use of permutation methods.

"I'm amazed that anybody is still doing permutation tests," wrote the anonymous reviewer. "There is probably nothing wrong technically with the paper, but I personally would reject it on grounds of irrelevance to current best statistical practice." To which the editor sought fit to add, "The reviewer is interested in estimation of interaction or main effects in the more general semi-parametric models currently studied in the literature. It is well known that permutation tests preserve the significance level but that all they do is answer yes or no."[1]

---

[1] A double untruth. First, permutation tests also yield interval estimates; see, for example, Garthwaite [1996]. Second, semiparametric methods are not appropriate for use with small-sample experimental designs, the topic of the submission.

Common Errors in Statistics (and How to Avoid Them), Third Edition. Edited by P. I. Good and J. W. Hardin
Copyright © 2009 John Wiley & Sons, Inc.

But one methodology can never be better than another, nor can estimation replace hypothesis testing or vice versa. Every methodology has a proper domain of application and another set of applications for which it fails. Every methodology has its drawbacks and its advantages, its assumptions and its sources of error. Let us seek the best from each statistical procedure.

The balance of this chapter is devoted to exposing the frailties of four of the "new" (and revived) techniques: bootstrap, Bayesian methods, meta-analysis, and permutation tests.

BOOTSTRAP

Many of the procedures discussed in this chapter fall victim to the erroneous perception that one can get more out of a sample or a series of samples than one actually puts in. One bootstrap expert learned that he was being considered for a position because management felt, "your knowledge of the bootstrap will help us to reduce the cost of sampling."

Michael Chernick, author of *Bootstrap Methods: A Practitioner's Guide* [2007], has documented six myths concerning the bootstrap:

1. Allows you to reduce your sample size requirements by replacing real data with simulated data—Not.
2. Allows you to stop thinking about your problem, the statistical design and probability model—Not.
3. No assumptions necessary—Not.
4. Can be applied to any problem—Not.
5. Only works asymptotically—Necessary sample size depends on the context.
6. Yields exact significance levels—Never.

Of course, the bootstrap does have many practical applications as witness its appearance in six of the chapters in this text.[2]

- Confidence intervals for population functionals that rely primarily on the center of the distribution such as the Mean, Median, and 40th through 60th percentiles.
- Model validation (see Appendix B)
- Estimating bias
- When all else fails
    - Behrans–Fisher problem (Good, 2005; Section 3.6.4)
    - Missing cells from an experimental design (Good, 2006; Section 5.6)
    - Sample-size determination

---

[2]If you're counting, we meet the bootstrap again in Chapters 10 and 11.

Limitations

As always, to use the bootstrap or any other statistical methodology effectively, one has to be aware of its limitations. The bootstrap is of value in any situation in which the sample can serve as a surrogate for the population.

If the sample is not representative of the population because the sample is small or biased, not selected at random, or its constituents are not independent of one another, then the bootstrap will fail.

Canty et al. [2006] also list data outliers, inconsistency of the bootstrap method, incorrect resampling model, wrong or inappropriate choice of statistic, nonpivotal test statistics, nonlinearity of the test statistic, and discreteness of the resample statistic as potential sources of error.

One of the first proposed uses of the bootstrap, illustrated in Chapter 3, was in providing an interval estimate for the sample median. Because the median or 50th percentile is in the center of the sample, virtually every element of the sample contributes to its determination. As we move out into the tails of a distribution, to determine the 20th percentile or the 90th, fewer and fewer elements of the sample are of assistance in making the estimate.

For a given size sample, bootstrap estimates of percentiles in the tails will always be less accurate than estimates of more centrally located percentiles. Similarly bootstrap interval estimates for the variance of a distribution will always be less accurate than estimates of central location such as the mean or median, as the variance depends strongly upon extreme values in the population.

One proposed remedy is the tilted bootstrap,[3] in which, instead of sampling each element of the original sample with equal probability, we weight the probabilities of selection so as to favor or discourage the selection of extreme values.

If we know something about the population distribution in advance—for example, if we know that the distribution is symmetric, or that it is chi-square with six degrees of freedom—then we may be able to take advantage of a parametric or semiparametric bootstrap as described in Chapter 5 Recognize that in doing so, you run the risk of introducing error through an inappropriate choice of parametric framework.

Problems due to the discreteness of the bootstrap statistic are usually evident from plots of bootstrap output. They can be addressed using a smooth bootstrap, as described in Davison and Hinkley [1997, Section 3.4].

BAYESIAN METHODOLOGY

Since being communicated to the Royal Society in 1763 by Reverend Thomas Bayes,[4] the eponymous Bayes' Theorem has exerted a near-fatal attraction on those exposed to it.[5] Much as a bell placed on the cat would magically resolve so many of the problems of the average house mouse, Bayes' straightforward, easily grasped mathematical

---

[3]See, for example, Hinkley and Shi [1989] and Phipps [1997].
[4]*Phil. Tran.* 1763; 53:376–398. Reproduced in: *Biometrika* 1958; 45: 293–315.
[5]The interested reader is directed to Keynes [1921] and Redmayne [1998] for some accounts.

formula would appear to provide the long-awaited basis for a robotic judge that is free of human prejudice.

On the plus side, Bayes' Theorem offers three main advantages:

1. It simplifies the combination of many different kinds of evidence, lab tests, animal experiments, and clinical trials and serves as an effective aid to decision making.
2. It permits evaluation of evidence in favor of a null hypothesis. In addition, with very large samples, a null hypothesis is not automatically rejected.
3. It provides dynamic flexibility *during* an experiment; sample sizes can be modified, measuring devices altered, subject populations changed, and endpoints redefined.

Suppose we have in hand a set of evidence $E = \{E_1, E_2, \ldots, E_n\}$ and thus have determined the conditional probability $\Pr\{A|E\}$ that some event A is true. A might be the event that O. J. Simpson killed his ex-wife, that the captain of the *Exxon Valdez* behaved recklessly, or some other incident whose truth or falsehood we wish to establish. An additional piece of evidence E_{n+1} now comes to light. Bayes' Theorem tell us that

$$\Pr\{A|E_1, \ldots, E_n, E_{n+1}\}$$

$$= \frac{\Pr\{E_{n+1}|A\}\Pr\{A|E_1, \ldots, E_n\}}{\Pr\{E_{n+1}|A\}\Pr\{A|E_1, \ldots, E_n\} + \Pr\{E_{n+1}| \sim A\}\Pr\{\sim A|E_1, \ldots, E_n\}}$$

where $\sim A$ (read "not A") is the event that A did not occur. Recall that $\Pr\{A\} + \Pr\{\sim A\} = 1$. $\Pr\{A|E_1, \ldots, E_n\}$ is the *prior* probability of A and $\Pr\{A|E_1, \ldots, E_n, E_{n+1}\}$ is the *posterior* probability of A once the item of evidence E_{n+1} is in hand. Gather sufficient evidence and we shall have an automatic verdict.

The problem with the application of Bayes' Theorem in practice comes at the beginning, when we have no evidence in hand and $n = 0$. What is the prior probability of A then?

Applications in the Courtroom[6]

Bayes' Theorem has seen little use in criminal trials because ultimately the theorem relies on unproven estimates rather than known facts.[7] Tribe [1971] states several objections, including the argument that a jury might use the evidence twice, once in its initial assessment of guilt—that is, to determine a prior probability—and a second time when the jury applies Bayes' Theorem. A further objection to the theorem's application is that if a man is innocent until proven guilty, the prior probability of his guilt must be zero; by Bayes' Theorem, the posterior probability of his guilt is

---

[6]The majority of this section is reprinted with permission from *Applying Statistics in the Courtroom*, by Phillip Good, Copyright 2001 by CRC Press Inc.

[7]See, for example, *People v Collins*, 68 Cal .2d 319, 36 ALR3d 1176 (1968).

also zero, rendering a trial unnecessary. The courts of several states have remained unmoved by this argument.[8]

In *State v. Spann*,[9] showing that the defendant had fathered the victim's child was key to establishing a charge of sexual assault. The state's expert testified that only 1% of the presumed relevant population of possible fathers had the type of blood and tissue that the father had and, further, that the defendant was included in that 1%. In other words, 99% of the male population at large was excluded. Next, the expert used Bayes' Theorem to show that the defendant had a posterior probability of fathering the victim's child of 96.5%.

> The expert testifying that the probability of defendant's paternity was 96.5% knew absolutely nothing about the facts of the case other than those revealed by blood and tissues tests of defendant, the victim, and the child. . . .[10]

> In calculating a final probability of paternity percentage, the expert relied in part on this 99% probability of exclusion. She also relied on an assumption of a 50% prior probability that defendant was the father. This assumption [was] not based on her knowledge of any evidence whatsoever in this case. . . . [She stated], "everything is equal . . . he may or may not be the father of the child."[11]

> Was the expert's opinion valid even if the jury disagreed with the assumption of 0.5 [50%]? If the jury concluded that the prior probability is 0.4 or 0.6, for example, the testimony gave them no idea of the consequences, no knowledge of what the impact (of such a change in the prior probability) would be on the formula that led to the ultimate opinion of the probability of paternity."[12]

> . . . [T]he expert's testimony should be required to include an explanation to the jury of what the probability of paternity would be for a varying range of such prior probabilities, running for example, from 0.1 to 0.9."[13]

In other words, Bayes' Theorem might prove applicable if, regardless of the form of the *a priori* distribution, one came to more or less the same conclusion.

Courts in California,[14] Illinois, Massachusetts,[15] Utah,[16] and Virginia[17] also have challenged the use of the 50:50 assumption. In *State v. Jackson*,[18] the expert did include a range of prior probabilities in her testimony, but the court ruled that the trial judge had erred in allowing the expert to testify as to the conclusions of Bayes'

---

[8]See, for example, *Davis v. State*, 476 N.E.2d 127 (Ind.App.1985), and *Griffith v. State of Texas*, 976 S.W.2d 241 (1998).
[9]130 N.J. 484 (1993).
[10]Id. 489.
[11]Id. 492.
[12]Id. 498.
[13]Id. 499.
[14]*State v. Jackson*, 320 NC 452, 358 S.E.2d 679 (1987).
[15]*Commonwealth v. Beausoleil*, 397 Mass. 206 (1986).
[16]*Kofford v. Flora*, 744 P.2d 1343, 1351–2 (1987).
[17]*Bridgeman v. Commonwealth*, 3 Va. App 523 (1986).
[18]320 N.C. 452 (1987).

Theorem in stating a conclusion that the defendant was "probably" the father of the victim's child.

In *Cole v. Cole*,[19] a civil action, the court rejected the admission of an expert's testimony of a high probability of paternity derived via Bayes' formula because there was strong evidence that the defendant was sterile as a result of a vasectomy.

> The source of much controversy is the statistical formula generally used to calculate the provability of paternity: Bayes' Theorem. Briefly, Bayes' Theorem shows how new statistical information alters a previously established probability. . . . When a laboratory uses Bayes' Theorem to calculate a probability of paternity it must first calculate a "prior probability of paternity." . . . This prior probability usually has no connection to the case at hand. Sometimes it reflects the previous success of the laboratory at excluding false fathers. Traditionally, laboratories use the figure 50%, which may or may not be appropriate in a given case.

> Critics suggest that this prior probability should take into account the circumstances of the particular case. For example, if the woman has accused three men of fathering her child or if there are reasons to doubt her credibility, or if there is evidence that the husband is infertile, as in the present case, then the prior probability should be reduced to less than 50%.[20]

The question remains as to what value to assign the prior probability. Another question is whether, absent sufficient knowledge to pin down the prior probability with any accuracy, we can make use of Bayes' Theorem at all. At trial, an expert called by the prosecution in *Plemel v. Walter*[21] used Bayes' Theorem to derive the probability of paternity.

> If the paternity index or its equivalents are presented as the probability of paternity, this amounts to an unstated assumption of a prior probability of 50 percent. . . . [T]he paternity index will equal the probability of paternity only when the other evidence in this case establishes prior odds of paternity of exactly one.[22]

> . . . [T]he expert is unqualified to state that any single figure is the accused's "probability of paternity." As noted above, such a statement requires an estimation of the strength of other evidence presented in the case (i.e., an estimation of the "prior probability of paternity"), an estimation that the expert is [in] no better position to make than the trier of fact.[23]

> Studies in Poland and New York City have suggested that this assumption [a 50 percent prior probability] favors the putative father because in an estimated 60 to 70 percent of paternity cases the mother's accusation of paternity is correct. Of course, the purpose of paternity litigation is to determine whether the mother's accusation is correct and for that reason it would be both unfair and improper to apply the assumption in any particular case.[24]

[19] 74 N.C.App. 247, aff'd. 314 N.C. 660 (1985).
[20] Id. 328.
[21] 303 Or. 262 (1987).
[22] Id. 272.
[23] Id. 275.
[24] Id. 276, fn. 9.

A remedy proposed by the Court is of interest to us:

> If the expert testifies to the defendant's paternity index or a substantially equivalent statistic, the expert must, if requested, calculate the probability that the defendant is the father by using more than a single assumption about the strength of the other evidence in the case. ... If the expert uses various assumptions and makes these assumptions known, the fact finder's attention will be directed to the other evidence in the case, and will not be misled into adopting the expert's assumption as to the correct weight assigned the other evidence. The expert should present calculations based on assumed prior probabilities of 0, 10, 20, ... , 90 and 100 percent.[25]

The courts of many other states have followed *Plemel*. "The better practice may be for the expert to testify to a range of prior probabilities, such as 10, 50 and 90 percent, and allow the trier of fact to determine which to use."[26]

Applications to Experiments and Clinical Trials

Outside the courtroom, where the rules of evidence are less rigorous, we have much greater latitude in the adoption of *a priori* distributions for the unknown parameter(s). Two approaches are common:

1. Adopting some synthetic distribution—a normal or a beta.
2. Using subjective probabilities.

The synthetic approach, though common among the more computational, is difficult to justify. The theoretical basis for an observation having a normal distribution is well known—the observation will be the sum of a large number of factors, each of which makes only a minute contribution to the total. But could such a description be applicable to a population parameter?

Here is an example of this approach taken from a report by D. A. Berry[27]:

> A study reported by Freireich et al.[28] was designed to evaluate the effectiveness of a chemotherapeutic agent, 6-mercaptopurine (6-MP), for the treatment of acute leukemia. Patients were randomized to therapy in pairs. Let p be the population proportion of pairs in which the 6-MP patient stays in remission longer than the placebo patient. (To distinguish probability p from a probability distribution concerning p, I will call it a population proportion or a propensity.) The null hypothesis H_0 is $p = 1/2$: no effect

---

[25]Id. 279. See, also Kaye [1988].

[26]*County of El Dorado v. Misura*, 33 Cal. App.4th 73 (1995), citing *Plemel*, supra, at p. 1219; Peterson (1982 at p. 691, fn. 74), *Paternity of M.J.B.* 144 Wis.2d 638, 643; *State v. Jackson*, 320 N.C.452, 455 (1987), and *Kammer v. Young*, 73 Md. App. 565, 571 (1988). See, also *State v. Spann*, 130 N.J. 484 at p. 499 (1993).

[27]The full report, titled "Using a Bayesian Approach in Medical Device Development," may be obtained from Donald A. Berry at the Institute of Statistics and Decision Sciences and Comprehensive Cancer Center, Duke University, Durham, NC 27708.

[28]*Blood* 1963; **21**:699–716.

of 6-MP. Let H_1 stand for the alternative hypothesis that $p > 1/2$. There were 21 pairs of patients in the study, and 18 of them favored 6-MP.

Suppose that the prior probability of the null hypothesis is 70 percent and that the remaining probability of 30 percent is on the interval $(0,1)$ uniformly. . . . So under the alternative hypothesis H_1, p has a uniform$(0,1)$ distribution. This is a mixture prior in the sense that it is 70 percent discrete and 30 percent continuous.

The uniform$(0,1)$ distribution is also the $\beta(1,1)$ distribution. Updating the $\beta(a,b)$ distribution after s successes and f failures is easy, namely, the new distribution is $\beta(a + s, b + f)$. So for $s = 18$ and $f = 3$, the posterior distribution under H_1 is $\beta(19,4)$.

The subjective approach places an added burden on the experimenter. As always, she must specify each of the following:

- Maximum acceptable frequency of Type I errors (that is, the significance level).
- Alternative hypotheses of interest.
- Power desired against each alternative.
- Losses associated with Type I and Type II errors.

With the Bayesian approach, she must also provide prior probabilities.

Arguing in favor of the use of subjective probabilities is that they permit incorporation of expert judgment in a formal way into inferences and decision making. Arguing against them, in the words of the late Edward Barankin, "How are you planning to get these values—beat them out of the researcher?" More appealing, if perhaps no more successful, approaches are described by Good [1950] and Kadane et al. [1980].

Bayes' Factor

An approach that allows us to take advantage of the opportunities Bayes' Theorem provides while avoiding its limitations and the objections raised in the courts is through the use of the minimum Bayes' factor.

In the words of Steven Goodman [2001]:

The odds we put on the null hypothesis (relative to others) using data external to a study is called the "prior odds," and the odds after seeing the data are the "posterior odds." The Bayes' factor tells us how far apart those odds are, that is, the degree to which the data from a study move us from our initial position. It is quite literally an epistemic odds ratio, the ratio of posterior to prior odds, although it is calculable from the data, without those odds. It is the ratio of the data's probability under two competing hypotheses.[29]

If we have a Bayes' factor equal to $1/10$ for the null hypothesis relative to the alternative hypothesis, it means that these study results have decreased the relative odds of the null hypothesis by 10-fold. For example, if the initial odds of the null were 1 (i.e., a probability of 50%), then the odds after the study would be $1/10$ (a probability of 9%). Suppose that the probability of the null hypothesis is high to begin with (as [it] typically [is] in data

[29]See Goodman [1999] and Greenland [1998].

dredging settings), say an odds of 9 (90%). Then a 10-fold decrease would change the odds of the null hypothesis to $9/10$ (a probability of 47%), still quite probable.

The appeal of the minimum Bayes' factor[30] is that it is calculated from the same information that goes into the P-value, and can easily be derived from standard analytic results, as described below. Quantitatively, it is only a small step from the P-value (and shares the liability of confounding the effect size with its precision).

The calculation [of the minimum Bayes' factor] goes like this. If a statistical test is based on a Gaussian approximation, the strongest Bayes' factor against the null hypothesis is $\exp(-Z^2/2)$, where Z is the number of standard errors from the null value. If the log-likelihood of a model is reported, the minimum Bayes' factor is simply the exponential of the difference between the log-likelihoods of two competing models (i.e., the ratio of their maximum likelihoods).

The minimum Bayes' factor described above does not involve a prior probability distribution over non-null hypotheses; it is a global minimum for all prior distributions. However, there is also a simple formula for the minimum Bayes' factor in the situation where the prior probability distribution is symmetric and descending around the null value. This is $-ep\ln(p)$,[31] where p is the fixed-sample size P-value. Table B.1 [not shown] shows the correspondence between P-values, Z- (or t-) scores, and the two forms of minimum Bayes' factors described above. Note that even the strongest evidence against the null hypothesis does not lower its odds as much as the P-value magnitude might lead people to believe. More importantly, the minimum Bayes' factor makes it clear that we cannot estimate the credibility of the null hypothesis without considering evidence outside the study.

Reading from Table B.1, a P-value of 0.01 represents a "weight of evidence" for the null hypothesis of somewhere between $1/25$ (0.04) and $1/8$ (0.13). In other words, the relative odds of the null hypothesis versus any alternative are at most $8-25$ times lower than they were before the study. If I am going to make a claim that a null effect is highly unlikely (e.g., less than 5%), it follows that I should have evidence outside the study that the prior probability of the null was no greater than 60%. If the relationship being studied is far-fetched (e.g., the probability of the null was greater than 60%), the evidence may still be too weak to make a strong knowledge claim. Conversely, even weak evidence in support of a highly plausible relationship may be enough for an author to make a convincing case.[32]

Two caveats:

1. Bayesian methods cannot be used in support of after-the-fact hypotheses for, by definition, an after-the-fact hypothesis has zero *a priori* probability and, thus, by Bayes' rule, zero *a posteriori* probability.

2. One hypothesis proving to be of greater predictive value than another in a given instance may be suggestive but is far from definitive in the absence of collateral evidence and proof of causal mechanisms. See, for example, Hodges [1987].

---

[30] As introduced by Edwards et al. [1963].
[31] See Bayarri and Berger [1998] and Berger and Selike [1987].
[32] Reprinted with permission from Lippincott Williams & Wilkins.

> ## WHEN USING BAYESIAN METHODS
>
> Do not use an arbitrary prior.
> Never report a *p*-value.
> Incorporate potential losses in the decision.
> Report the Bayes' factor.

META-ANALYSIS

Meta-analysis is a set of techniques that allow us to combine the results of a series of small trials and observational studies. With the appropriate meta-analysis, we can, in theory, obtain more precise estimates of main effects, test *a priori* hypotheses about subgroups, and determine the number of observations needed for large-scale randomized trials.

By putting together all available data, meta-analyses are also better placed than individual trials to answer questions about whether an overall study result varies among subgroups—for example, among men and women, older and younger patients, or subjects with different degrees of severity of disease.

Meta-analysis should be viewed as an observational study of the evidence. The steps involved are similar to [those of] any other research undertaking: formulation of the problem to be addressed, collection and analysis of the data, and reporting of the results. Researchers should write in advance a detailed research protocol that clearly states the objectives, the hypotheses to be tested, the subgroups of interest, and the proposed methods and criteria for identifying and selecting relevant studies and extracting and analysing information.

—Egger, Smith, and Phillips [1997][33]

Too many studies end with inconclusive results because of the relatively small number of observations that were made. The researcher can't quite reject the null hypothesis but isn't quite ready to embrace the null hypothesis either. As we saw in Chapter 1, a post hoc subgroup analysis can suggest an additional relationship, but the relationship cannot be subject to statistical test in the absence of additional data.

In performing a meta-analysis, we need to distinguish between observational studies and randomized trials.

Confounding and selection bias can easily distort the findings from observational studies. Egger et al. [1998]:

An important criterion supporting causality of associations is a dose-response relation. In occupational epidemiology the quest to show such an association can lead to very different groups of employees being compared. In a meta-analysis that examined the link between exposure to formaldehyde and cancer, funeral directors and embalmers

[33]Reprinted with permission from the BMJ Publishing Group.

(high exposure) were compared with anatomists and pathologists (intermediate to high exposure) and with industrial workers (low to high exposure, depending on job assignment). There is a striking deficit of deaths from lung cancer among anatomists and pathologists [standardized mortality ratio 33 (95% confidence interval 22 to 47)], which is most likely to be due to a lower prevalence of smoking among this group. In this situation few would argue that formaldehyde protects against lung cancer. In other instances, however, such selection bias may be less obvious.[34]

On the other hand, much may be gained by a careful examination of possible sources of heterogeneity between the results from observational studies.

Publication and selection bias also plague the meta-analysis of completely randomized trials. Inconclusive or negative results seldom appear in print [Götzsche, 1987; Begg and Berlin, 1988; Chalmers et al., 1990; Eastebrook et al., 1991] and are unlikely even to be submitted for publication. One can't incorporate in a meta-analysis what one doesn't know about.

"Authors who try to evaluate the quality of randomized trials, possibly for the purpose of weighting them in meta-analyses, need to . . . concern themselves also with the restrictions on the randomization and the extent to which compromised allocation concealment led to selection bias" [Berger, 2006]. Similarly, the decision as to which studies to incorporate can dramatically affect the results. Meta-analyses of the same issue may reach opposite conclusions, as shown by assessments of low molecular weight heparin in the prevention of perioperative thrombosis [Leizorovicz et al., 1992; Nurmohamed et al., 1992] and of second-line antirheumatic drugs in the treatment of rheumatoid arthritis [Felson et al., 1990; Götzsche et al., 1992]. Meta-analyses showing the benefit of statistical significance and clinical importance have been contradicted later by large randomized trials [Egger et al., 1997].

Where there are substantial differences between the different studies incorporated in a meta-analysis (their subjects or their environments) or substantial quantitative differences in the results from the different trials, a single overall summary estimate of treatment benefit has little practical applicability [Horwitz, 1995]. Any analysis that ignores this heterogeneity is clinically misleading and scientifically naive [Thompson, 1994]. Heterogeneity should be scrutinized, with an attempt to explain it [Bailey, 1987; Chalmers, 1991; Berkey et al., 1995; Victor, 1995].

Bayesian Methods

Bayesian methods can be effective in meta-analyses; see, for example, Mosteller and Chalmers [1992]. In such situations, the parameters of various trials are considered to be random samples from a distribution of trial parameters. The parameters of this higher-level distribution are called hyperparameters, and they also have distributions. The model is called hierarchical. The extent to which the various trials reinforce each other is determined by the data. If the trials are very similar, the variation of the hyperparameters will be small, and the analysis will be very close to a classical

---

[34]Reprinted with permission from the BMJ Publishing Group.

meta-analysis. If the trials do not reinforce each other, the conclusions of the hierarchical Bayesian analysis will show a very high variance in the results.

A hierarchical Bayesian analysis avoids the necessity of a prior decision as to
whether the trials can be combined; the extent of the combination is determined
purely by the data. This does not come for free; in contrast to the meta-analyses discussed above, all the original data (or at least the sufficient statistics) must be available
for inclusion in the hierarchical model. The Bayesian method is also vulnerable to all
the selection bias issues discussed above.

Guidelines for a Meta-Analysis

- A detailed research protocol for the meta-analysis should be prepared in advance.
  The criteria for inclusion and the statistical method employed should be docu-
  mented in the materials and methods section of the subsequent report.
- Meta-analysis should be restricted to randomized, controlled trials.
- Heterogeneity in the trial results should be documented and explained.
- Do not attempt to compare treatments investigated in unrelated trials. (Suppose,
  as a counterexample, that Old was given, as always, to low-risk patients in one set
  of trials, while New was given to high-risk patients in another.)
- Individual patient data, rather than published summary statistics, often are
  required for meaningful subgroup analyses. This is a major reason why we
  favor the modern trend of journals to insist that all data reported on in their
  pages be made available by website to all investigators.
- Johann Kepler was able to formulate his laws only because (1) Tycho Brahe
  had made over 30 years of precise (for the time) astronomical observations
  and (2) Kepler married Brahe's daughter and thus gained access to his data.

PERMUTATION TESTS

Permutation tests first introduced by Pitman [1937, 1938] are often lauded erroneously
in the literature as assumption-free panaceas. Nothing could be further from the truth.

Permutation tests yield exact significance levels only if the labels on the observations are weakly exchangeable under the null hypothesis. Thus, they cannot be
successfully applied to the coefficients in a multivariate regression.

On the other hand, if the observations are weakly exchangeable under the null
hypothesis,[35] then permutation tests are the method of choice for the following:

- Two-sample multivariate comparisons.
- Crossover designs.
- k-Sample comparisons.
- Type I censoring.
- Contingency tables whenever there are 12 or fewer observations in each
  subsample.

[35]The concept is made precise in Good [2002].

Moreover, permutation methods can be used both to test hypotheses and to obtain interval estimates of parameters.

In other practical situations, such as the two-sample comparison of means (crossover designs being the exception) and bivariate correlation, permutation tests offer no advantage over parametric methods like Students' t and Pearson's correlation.

By making use of the permutation distribution of a test statistic, one is no longer limited by the availability of tables, but is always free to employ the most powerful statistic against the alternative(s) of interest or the statistic that will be most effective in minimizing the losses of interest.

For example, for comparing the means of several populations, one may use any of the following statistics:

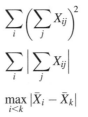

$$\sum_i \left(\sum_j X_{ij} \right)^2$$

$$\sum_i \left| \sum_j X_{ij} \right|$$

$$\max_{i<k} |\bar{X}_i - \bar{X}_k|$$

Permutation methods can be applied to the original observations, to the ranks of the observations (when they are known as rank tests), or to normal and distribution scores.

Permutation tests are often described as "analyzing an experiment in the way it was designed," see, for example, Bradley [1968]. But if the design is flawed, then so will the analysis be. In a sidebar in Chapter 3, we described a flawed crossover experiment in which subjects were assigned at random with replacement to treatment sequence, so that the final design was severely unbalanced. Nonetheless, the design might have been analyzed correctly by permutation means, had the designer not chosen to "analyze the experiment in the way it was designed." Specifically, the patients' data used in the primary analysis was reassigned to the 18 treatment sequences randomly, using $1/18$ as the probability within each randomization block. The resultant test was an inexact bootstrap rather than an exact permutation.

TO LEARN MORE

Potential flaws in the bootstrap approach are considered by Schenker [1985], Wu [1986], Diciccio and Romano [1988], Efron [1988, 1992], Knight [1989], and Gine and Zinn [1989]. Some improvements are suggested by Fisher and Hall [1990, 1991]. Canty et al. [2006] provide a set of diagnostics for detecting and dealing with potential error sources.

Berry and Stangl [1996] include a collection of case studies in Bayesian biostatistics. Clemen, Jones, and Winkler subject Bayesian methods to an empirical evaluation. Kass and Raferty [1995] discuss the problem of establishing priors along with a set of practical examples. The Bayes' factor can be used as a test statistic; see Good [1992].

For more on the strengths and limitations of meta-analysis, see Teagarden [1989], Gurevitch and Hedges [1993], Horwitz [1995], Egger and Smith [1997], Egger,

Smith, and Phillips [1997], Smith, Egger, and Phillips [1997], Smith and Egger [1998], Smeeth, Haines, and Ebrahim [1999], and Gillett [2001]. To learn about the appropriate statistical procedures, see Adams, Gurevitch, and Rosenberg [1997], Berlin et al. [1989], Hedges and Olkin [1985]. Sharp and Thompson [1996, 2000] analyze the relationship between treatment benefit and underlying risk. Smith, Spiegelhalter, and Parmar describe a Bayesian meta-analysis.

For practical, worked-through examples of hierarchical Bayesian analysis, see Palmer, Graham, White and Hansen [1998], Harley and Myers [2001], and Su, Adkison, and Van Alen [2001]. Theoretical development may be found in Mosteller and Chalmers [1992] and Carlin and Louis [1996].

The lack of access to the raw data underlying published studies is a matter of ongoing concern as the conclusions of meta-analyses based on published results may differ substantially from those based on all available evidence; see Simes [1986], Stewart and Parmar [1993], Moher et al. [1999], Eysenbach and Sa [2001], and Hutchon [2001].

Permutation methods and their applications are described in Pesarin [1990, 2001], Manley [1997], Mielke and Berry [2001], and Good [2005]. For a description of some robust permutation tests, see Lambert [1985] and Maritz [1996]. Berger [2000] reviews the pros and cons of permutation tests.

PART III

REPORTS

8

REPORTING YOUR RESULTS

Cut out the appropriate part of the computer output and paste it onto the draft of the paper.
—Dyke (tongue in cheek) [1997]

The focus of this chapter is on what to report and how to report it. Reportable elements include the experimental design and its objectives, its analysis, and the sources and amounts of missing data. Guidelines for table construction are provided. The bootstrap is proposed as an alternative to the standard error as a measure of precision. The value and limitations of *p*-values and confidence intervals are summarized. Practical significance is distinguished from statistical significance and induction from deduction.

FUNDAMENTALS

Few experimenters fail to list number of subjects, doses administered, and dose intervals in their reports. But many fail to provide the details of power and sample size calculations. Feng et al. [2001] found that such careless investigators also report a higher proportion of nonsignificant intervention effects, indicating underpowered studies. Your report should include all the estimates you used in establishing sample sizes, along with the smallest effect of practical interest that you hoped to detect and the corresponding power.

Too often, inadequate attention is given to describing treatment allocation and the ones who got away. We consider both topics in what follows.

Common Errors in Statistics (and How to Avoid Them), Third Edition. Edited by P. I. Good and J. W. Hardin
Copyright © 2009 John Wiley & Sons, Inc.

Treatment Allocation[1]

Allocation details should be fully described in your reports, including dictated allocation versus allocation discretion, randomization, advance preparation of the allocation sequence, allocation concealment, fixed versus varying allocation proportions, restricted randomization, masking, simultaneous versus sequential randomization, enrollment discretion, and the possibility of intent to treat.

Allocation discretion may be available to the investigator, the patient, both, or neither (dictated allocation). Were investigators permitted to assign treatment based on patient characteristics? Could patients select their own treatment from among a given set of choices?

Was actual (not virtual, quasi-, or pseudo-) randomization employed? Was the allocation sequence predictable? (For example, patients with even accession numbers or patients with odd accession numbers receive the active treatment; the others receive the control.)

Was randomization *conventional*, that is, was the allocation sequence generated in advance of screening any patients?

Was allocation concealed prior to its execution? As Vance W. Berger and Costas A. Christophi relate in a personal communication, "This is not itself a reportable design feature, so a claim of allocation concealment should be accompanied by specific design features. For example, one may conceal the allocation sequence; and instead of using envelopes, patient enrollment may involve calling the baseline information of the patient to be enrolled in to a central number to receive the allocation."

Was randomization restricted or unrestricted? Randomization is *unrestricted* if a patient's likelihood of receiving either treatment is independent of all previous allocations and is *restricted* otherwise. If both treatment groups must be assigned equally often, then prior allocations determine the final ones. Were the proportions also hidden?

Were treatment codes concealed until all patients were randomized and the database was locked? Were there instances of codes being revealed accidentally? Senn [1995] warns, "Investigators should delude neither themselves, nor those who read their results, into believing that simply because some aspects of their trial were double-blind that therefore all the virtues of such trials apply to all their conclusions." Masking can rarely, if ever, be assured; see also Day [1998].

Was randomization simultaneous, block simultaneous, or *sequential*? A blocked randomization is *block simultaneous* if all patients within any given block are identified and assigned accession numbers prior to any patient in that block being treated.

And, not least, was intent to treat permitted?

---

[1]The material in this section relies heavily on a personal communication from Vance W. Berger and Costas A. Christophi.

Adequacy of Blinding

> The current lack of reporting on the success of blinding provides little evidence that success of blinding is maintained in placebo controlled trials. Trialists and editors should make a concerted effort to incorporate, report, and publish such information and its potential effect on study results.
>
> —Fergusson et al. [2004]

We, too, believe authors should add a section describing their assessment of blinding to all their reports. Here is an example taken from Turner et al. [2005]:

> The adequacy of the study's blinding procedures was assessed according to the subjects' responses when asked which study medication they believed they were taking ("active," "placebo," or "don't know"). This question was asked at the end of the prophylaxis phase just before virus challenge and again after administration of the third dose of study medication in the treatment phase of the trial.

A pilot study may be done without blinding as a prelude to more extensive controlled trials, but this lack should be made explicit in your report, as in Rozen et al. [2008]. Controls should always be employed, lest unforseen and unrelated events such as an epidemic yield a misleading result.

WRITE IN ORDINARY LANGUAGE

Use common terminology in preference to "statistics-speak." For example, write "We will graph" in preference to "We will graphically depict."

Roberts et al. [2007] often challenge the reader rather than inform. Here are some examples:

> We evaluated a series of hypotheses regarding an association between *in utero* residential "exposure" to specific agricultural pesticides (that is, maternal residence in close proximity to sites of application) and the development of ASD by linking existing databases using a retrospective case-control design.

> We operationalized the hypotheses of association between exposure and outcome based on known embryological phenomena.

> Temporal parameters were chosen to reflect the hypotheses that the periods immediately prior to and during Central Nervous System (CNS) embryogenesis, neural tube closure, and entire gestation could represent critical windows for exposure.

This last paragraph translates as "We divided the gestational period into three strata based upon the stage of CNS development in the fetus." (At least, we think that's what they meant.)

Finally, these authors make repeated mention of "the 4th non-zero quartile coefficient." We freely confess that we don't know what this is.

Missing Data[2]

Every experiment or survey has its exceptions. You must report the raw numbers of such exceptions and, in some instances, provide additional analyses that analyze or compensate for them. Typical exceptions include the following:

Did Not Participate. Subjects who were eligible and available but did not participate in the study—this group should be broken down further into those who were approached but chose not to participate and those who were not approached. With a mail-in survey, for example, we would distinguish between those whose envelopes were returned "address unknown" and those who simply did not reply.

Ineligibles. In some instances, circumstances may not permit deferring treatment until the subject's eligibility can be determined.

For example, an individual arrives at a study center in critical condition; the study protocol calls for a series of tests, the results of which may not be back for several days, but in the opinion of the examining physician treatment must begin immediately. The patient is randomized to treatment and only later is it determined that the patient is ineligible.

The solution is to present two forms of the final analysis, one incorporating all patients, the other limited to those who were actually eligible.

Withdrawals. Subjects who enrolled in the study but did not complete it including both dropouts and noncompliant patients. These patients might be subdivided further based on the point in the study at which they dropped out.

At issue is whether such withdrawals were treatment related or not. For example, the gastro-intestinal side effects associated with erythromycin are such that many patients (including both authors) may refuse to continue with the drug. Traditional statistical methods are not applicable when withdrawals are treatment related.

Crossovers. If the design provided for intent-to-treat, a noncompliant patient may still continue in the study after being reassigned to an alternate treatment. Two sets of results should be reported: the first for all patients who completed the trials (retaining their original treatment assignments for the purpose of analysis); the second restricted to the smaller number patients who persisted in the treatment groups to which they were originally assigned.

Missing Data. Missing data are common, expensive, and preventable in many instances.

The primary endpoint of a recent clinical study of various cardiovascular techniques was based on the analysis of follow-up angiograms. Although more than 750 patients were enrolled in the study, only 523 had the necessary angiograms.

---

[2]The material in this section is reprinted with permission from *Manager's Guide to Design and Conduct of Clinical Trials*, Phillip Good, Wiley, 2002.

Almost a third of the monies spent on the trials had been wasted. This result is not atypical. Capaldi and Patterson [1987] uncovered an average attrition rate of 47% in studies lasting 4 to 10 years.

You need to analyze the data to ensure that the proportions of missing observations are the same in all treatment groups. Again, traditional statistical methods are applicable only if missing data are not treatment related.

Deaths and disabling accidents and diseases, whether [or not] directly related to the condition being treated, are common in long-term trials in the elderly and high-risk populations. Or individuals are simply lost to sight ("no forwarding address") in highly mobile populations.

Lang and Secic [1997, p. 22] suggest a chart such as that depicted in Figure 3.5 as the most effective way to communicate all the information regarding missing data. Censored and off-scale measurements should be described separately and their numbers indicated in the corresponding tables.

WILL THE REAL n PLEASE STAND UP?

Fujita et al. [1995] describes an experiment in which 58 elderly hospitalized patients were divided into three groups at random. The number in each group was not reported. More important, this article omitted to say that at the end of the 30-month study, only 16 patients remained! Indeed, only 29 patients reported for the 12-month follow-up.

Fortunately, a follow-up report, Fujita et al. [1996], in a different journal supplied the missing values. Alas, the investigators persisted in comparing the mean baseline values of all patients entered in the study with the mean final values of the very few patients who completed it.

The study entailed the administration of various calcium supplements to a group of elderly individuals. The sickest, frailest individuals, and thus the ones with the lowest starting-baseline values, were almost certainly the ones who were lost to follow-up. But every time such a sick individual with a low baseline value dropped out of the study, the average for the group that remained rose of mathematical necessity.

The appropriate comparison is the within-individual changes of those who were in the study at the beginning and at the end. Alas, Fujita et al. did not include the original data in their articles, so the correct comparison is not possible.

A decade later, the same group of investigators, Fujita et al. [2004], published a third analysis of the same flawed study—this time omitting all mention of declining sample size and adding a series of misleading graphs using truncated vertical scales. Although the phrase "double blind" appears in the title of this article, readers were left to puzzle out how the double-blind aspect of the study was accomplished. Nor was there mention of blinding in the two previous articles reporting on this same study.

DESCRIPTIVE STATISTICS

In this section, we consider how to summarize your data most effectively, whether they comprise a sample or the entire population.

Binomial Trials

The most effective way of summarizing the results of a series of binomial trials is by recording the number of trials and the number of successes—for example, the number of coin flips and the number of heads, the number of patients treated and the number who got better, and so forth.

When trials can have three to five possible outcomes, the results are best presented in tabular form (as in Table 8.1) or in the form of a bar chart, whether the outcomes are ordered (no effect, small effect, large effect) or unordered (win, lose, tie). Both forms also provide for side-by-side comparisons of several sets of trials.

For the reasons discussed in the next chapter, we do *not* recommend the use of pie charts.

Categorical Data

When data fall into categories such as male versus female, black versus Hispanic versus oriental versus white, or in favor versus against versus undecided, we may display the results for a single categorical variable in the form of a bar chart. If there are multiple variables to be considered, the best way to display the results is in the form of a contingency table, as shown in Tables 8.2 and 8.3.

Note that Table 8.2 is a highly effective way of summarizing the data from nine different contingency tables similar to Table 8.3.

We can also summarize a single 2×2 table, like Table 8.3, simply by reporting the *odds ratio*, which takes the value $1 \times 12/(3 \times 18)$. In the more general case where a 2×2 table takes the form the odds ratio is $p(1 - s)/(1 - p)s$.

pn	$(1-p)n$
sm	$(1-s)m$

TABLE 8.1. RBIs per Game

	0	1	2	>2	Didn't Play
Good	2	3	2	1	3
Hardin	1	2	1	4	0

TABLE 8.2. Sandoz Drug Data

| Test Site | New Drug | | Control Drug | |
	Response	#	Response	#
1	0	15	0	15
2	0	39	6	32
3	1	20	3	18
4	1	14	2	15
5	1	20	2	19
6	0	12	2	10
7	3	49	10	42
8	0	19	2	17
9	1	14	0	15

TABLE 8.3. Sandoz Data, Site 3

	Response	No Response
New drug	1	19
Control	3	15

Rare Events

Reporting on events that are rare and random in time and/or space like suicides, accidental drownings, radioactive decay, the seeding of trees and thistles by the winds, and the sales of Dr. Good's novels[3] can be done in any of three different ways:

1. A statement as to the average interval between events—three days in the case of Dr. Good's novels.
2. A statement as to the average number of events per interval—10 per month in the case of Dr. Good's novels.
3. A listing in contingency table form of the frequency distribution of the events (see Table 8.1).

The clustering of random events is to be expected and *not* to be remarked upon. As a concrete example, while the physical laws which govern the universe are thought to be everywhere the same, the distribution of stars and galaxies is far from uniform; stars and galaxies are to be found everywhere in clusters and clusters of clusters [see Neyman and Scott, 1952].

---

[3]Search for "Sad and Angry Man" or "Luke Jackson," at http://amazon.com.

Measurements

Measurements such as weight, blood pressure, and lap time are normally made on a continuous or, more accurately, a *metric* scale. One can usefully talk about differences in measurements, such as the difference in mb Hg between one's blood pressure taken before and after smoking a cigarette. When a group of measurements are taken and a quick summary is desired, we can provide the arithmetic mean, the geometric mean, the median, the number of modes, or the percentiles of the observations' frequency distribution.

For one's own edification as opposed to a final report, one should begin by displaying some kind of frequency distribution. A box-and-whiskers plot (Fig. 8.1a) is superior to a dot plot (Fig. 8.1b), because it also tells us what the mean, median, and interquartile range of the data are. For small samples, the combined plot (Fig. 8.1c) may be the most informative.

Which Mean?

For small samples of three to five observations, summary statistics are virtually meaningless. Reproduce the actual observations; this is easier to do and more informative.

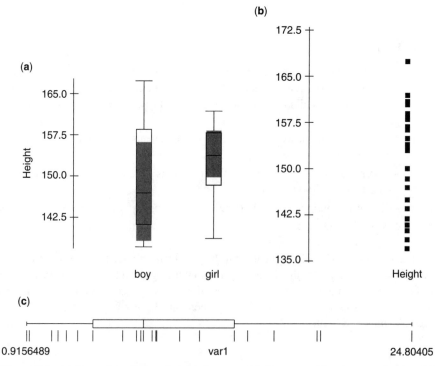

Figure 8.1. (a) Box plot of class heights by sex. (b) One-way strip chart or dotplot. (c) Combination boxplot (top section) and one-way strip chart.

Though the arithmetic mean or average is in common use for summarizing measurements, it can be very misleading. For example, the mean income in most countries is far in excess of the *median* income, or 50th percentile, to which most of us can relate. When the arithmetic mean is meaningful, it is usually equal to or close to the median. Consider reporting the median in the first place.

The *geometric mean* is more appropriate than the arithmetic in three sets of circumstances:

1. When losses or gains can best be expressed as a percentage rather than a fixed value.
2. When rapid growth is involved, as is the case with bacterial and viral populations.
3. When the data span several orders of magnitude, as with the concentration of pollutants.

The purpose of your inquiry must be kept in mind. The distribution of orders in dollars from a machinery plant is likely to be skewed by a few large orders. The median dollar value will be of interest in describing sales and appraising salespeople; the mean dollar value will be of interest in estimating revenues and profits.

Whether you report a mean or a median, be sure to report only a sensible number of decimal places. Most statistical packages can give you 9 or 10. Don't use them. If your observations were to the nearest integer, your report on the mean should include only a single decimal place. For guides to the appropriate number of digits, see Ehrenberg [1977] and, for percentages, van Belle [2002, Table 7.4].

Most populations are actually mixtures of populations. If multiple modes are observed in samples greater than 25 in size, the number of modes should be reported.

Ordinal Data

Ordinal data include measurements but also include observations that, while ordered, cannot be usefully added and subtracted, as measurements can. Observations recorded on the familiar Likert scale of 1—Disliked Intensely to 9—Liked Very Much, with 5 representing Indifference, are an example of ordinal but nonmetric data. One cannot assume that the difference between Disliked Intensely (1) and Disliked (3) is the same as between Disliked (3) and Indifferent (5). Thus, an arithmetic average or a variance would not be at all meaningful.

One can report such results in tabular form, in bar charts, or by providing key percentiles such as the minimum, median, and maximum.

Tables

Is text, a table, or a graph the best means of presenting results? Dyke [1997], argue "Tables with appropriate marginal means are often the best method of presenting results, occasionally replaced (or supplemented) by diagrams, usually graphs or histograms." Van Belle [2002] warns that aberrant values often can be more apparent in graphical form. An argument in favor of the use of ActivStats® for exploratory analysis is that one can so easily go back and forth from viewing the table to viewing the graph.

A sentence structure should be used for displaying two to five numbers, as in "The blood type of the population of the United States is approximately 45% O, 40% A, 11% B, and 4% AB." Note that the blood types are ordered by frequency.

Marginal means may be omitted only if they have already appeared in other tables.[4] Sample sizes should always be specified.

Among our own worst offenses is the failure to follow van Belle's [2002, p.154] advice to "Use the table heading to convey critical information. Do not stint. The more informative the heading, the better the table."

Consider adding a row (or column, or both) of contrasts; "for example, if the table has only two rows we could add a row of differences, row 1 minus row 2: if there are more than two rows, some other contrast might be useful, perhaps 'mean haploid minus mean diploid', or 'linear component of effect of N-fertilizer'."[4] Indicate the variability of these contrasts.

Tables dealing with two-factor arrays are straightforward, provided that confidence limits, least standard deviations, and standard errors are clearly associated with the correct set of figures. Tables involving three or more factors are not always immediately clear to the reader and are best avoided.

Are the results expressed in appropriate units? For example, are parts per thousand more natural in a specific case than percentages? Have we rounded off to the correct degree of precision, taking account of what we know about the variability of the results and considering whether they will be used by the reader, perhaps by multiplying by a constant factor, or by another variate, such as percent of dry matter?

Dyke [1997], also advises us that "Residuals should be tabulated and presented as part of routine analysis; any [statistical] package that does not offer this option was probably produced by someone out of touch with research workers, certainly with those working with field crops." Best of all is a display of residuals aligned in rows and columns as the plots were aligned in the field.

A table of residuals (or tables, if there are several strata) can alert us to the presence of outliers and may also reveal patterns in the data not considered previously.

THE WRONG WAY

In a two-factor experiment (litigated versus nonlitigated animals, AACa versus $CaCO_3$ dietary supplements), Tokita et al. [1993] studied rats in groups of sizes five, five, three, and three, respectively. The authors did not report their observations in tabular form. They reported a few of the standard deviations in a

---

[4]Dyke [1997]. Reprinted with permission from Elsevier Science.

summary, but only a few. Graphs were provided despite the paucity of observations; vertical bars accompanied the data points on the graphs, but the basis for their calculation was not provided. Although statistical significance was claimed, the statistical procedures used were not described.

Dispersion, Precision, and Accuracy

The terms dispersion, precision, and accuracy are often confused. Dispersion refers to the variation within a sample or a population. Standard measures of dispersion include the variance, the mean absolute deviation, the interquartile range, and the range.

Precision refers to how close several estimates based upon successive samples will come to one another. Accuracy refers to how close an estimate based on a sample will come to the population parameter it is estimating.

A satire of the Robin Hood legend depicts Robin splitting his first arrow with his second and then his second arrow with his third in a highly precise display of shooting. Then the camera pulls back and we see that all three arrows hit a nearby cow rather than the target. Precise, but highly inaccurate, shooting.

An individual confides on a statistics bulletin board that he is unsure how to get a confidence interval (a measure of the precision of an estimate) for census figures. If the census included or attempted to include all members of the population, the answer is, "You can't." One can complain of the inaccuracy of census figures (for the census surely excludes many homeless citizens) but not of the imprecision of figures based on a complete enumeration.

STANDARD ERROR

One of the most egregious errors in statistics, one encouraged, if not insisted upon by the editors of journals in the biological and social sciences, is the use of the notation "Mean $\pm$ Standard Error" to report the results of a set of observations.

The standard error is a useful measure of population dispersion *if* the observations are continuous measurements that come from a normal or Gaussian distribution. If the observations are normally distributed, as in the bell-shaped curve depicted in Figure 8.2, then in 95% of the samples we would expect the sample mean to lie within two standard errors of the mean of our original sample.

But if the observations come from a nonsymmetric distribution like an exponential or a Poisson, or a truncated distribution like the uniform, or a mixture of populations, we cannot draw any such inference.

Recall that the standard error equals the standard deviation divided by the square root of the sample size, $SD/\sqrt{n}$ or

$$\sum (x_i - \bar{x})^2 / \sqrt{n(n-1)}.$$

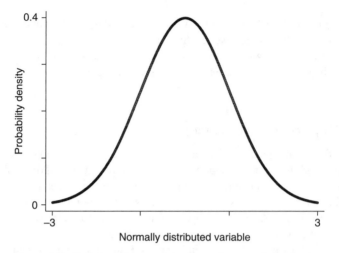

Figure 8.2. Bell-shaped symmetric curve of a normal distribution.

As the standard error depends on the squares of individual observations, it is particularly sensitive to outliers. A few extra large observations will have a dramatic impact on its value.

If you can't be sure that your observations come from a normal distribution, then consider reporting your results either in the form of a histogram, as in Figure 8.3a, or a box-and-whiskers plot, as in Figure 8.3b. See, also Lang and Secic [1997, p. 50].

If your objective is to report the precision of your estimate of the mean or median, then the standard error may be meaningful, provided that the mean of your observations is normally distributed.

The good news is that the sample mean often will have a normal distribution even when the observations do not come from a normal distribution. This is because the sum of a large number of random variables each of which makes only a small contribution to the total is a normally distributed random variable.[5] And in a sample mean based on n observations, each contributes only $1/n$th of the total. How close the fit is to a normal distribution will depend upon the size of the sample and the distribution from which the observations are drawn.

The distribution of a uniform random number $U[0,1]$ is a far cry from the bell-shaped curve of Figure 8.2. Only values between 0 and 1 have a positive probability, and in stark contrast to the normal distribution, no range of values between 0 and 1 is more likely than any other of the same length. The only element that the uniform and normal distributions have in common is that they are each symmetric about the population mean. Yet, to obtain normally distributed random numbers for use in simulations, we were once taught to generate 12 uniformly distributed random numbers and then take their average.

---

[5]This result is generally referred to as the Central Limit Theorem. Formal proof can be found in a number of textbooks, including Feller [1966, p. 253].

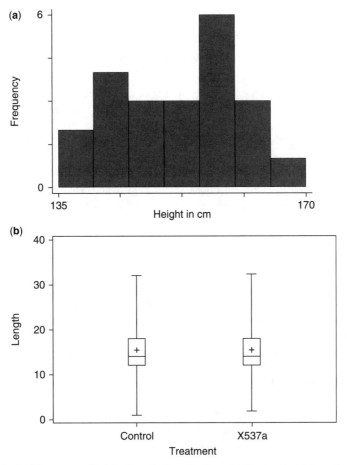

Figure 8.3. (a) Histogram of heights in a sixth-grade class. (b) Box-and-whiskers plot. The box encompasses the middle 50% of each sample, while the "whiskers" lead to the smallest and largest values. The line through the box is the median of the sample; that is, 50% of the sample is larger than this value, while 50% is smaller. The plus sign indicates the sample mean. Note that the mean is shifted in the direction of a small number of very large values.

Apparently, 12 is a large enough number for a sample mean to be normally distributed when the variables come from a uniform distribution. But take a smaller sample of observations from a $U[0,1]$ population and the distribution of its mean would look less like a bell-shaped curve.

A general rule of thumb is that the mean of a sample of 8 to 25 observations will have a distribution that is close enough to the normal for the standard error to be meaningful. The more nonsymmetric the original distribution, the larger the sample size required. At least 25 observations are needed for a binomial distribution with $p = 0.1$.

Even the mean of observations taken from a mixture of distributions (males and females, tall Zulu and short Bantu)—visualize a distribution curve resembling

| | |

142.25   Medians of Bootstrap Samples   158.25

Figure 8.4. Rugplot of 50 bootstrap medians derived from a sample of sixth graders' heights.

a camel with multiple humps—will have a normal distribution if the sample size is large enough. Of course, this mean (or even the median) conceals the fact that the sample was taken from a mixture of distributions.

If the underlying distribution is not symmetric, the use of the $\pm$ SE notation can be deceptive, as it suggests a nonexistent symmetry. For samples from nonsymmetric distributions of size six or less, tabulate the minimum, the median, and the maximum. For samples of size seven and up, consider using a box-and-whiskers plot. For samples of size 16 and up, the bootstrap, described in Chapters 5 and 6, may provide the answer you need.

As in those chapters, we would treat the original sample as a stand-in for the population and resample from it repeatedly, 1000 times or so, with replacement, computing the sample statistic each time to obtain a distribution similar to that depicted in Figure 8.4. To provide an interpretation compatible with that given the standard error when used with a sample from a normally distributed population, we would want to report the values of the 16th and 84th percentiles of the bootstrap distribution along with the sample statistic.

When the estimator is other than the mean, we cannot count on the Central Limit Theorem to ensure a symmetric sampling distribution. We recommend that you use the bootstrap whenever you report an estimate of a ratio or dispersion.

If you possess some prior knowledge of the shape of the population distribution, you should take advantage of that knowledge by using a parametric bootstrap (see Chapter 5). The parametric bootstrap is particularly recommended for use in determining the precision of percentiles in the tails (P_{20}, P_{10}, P_{90}, and so forth).

p-VALUES

> The p-value is *not* the probability that the null hypothesis is true.
>
> —Yoccoz [1991]

Before interpreting and commenting on p-values, it's well to remember that in contrast to the significance level, the p-value is a random variable that varies from sample to sample. There may be highly significant differences between two populations, and yet the samples taken from those populations and the resulting p-value may not reveal that difference. Consequently, it is not appropriate for us to compare the p-values from two distinct experiments, or from tests on two variables measured in the same experiment, and declare that one is more significant than the other.

If we agree in advance of examining the data that we will reject the hypothesis if the p-value is less than 5%, then our significance level is 5%. Whether our p-value proves to be 4.9% or 1% or 0.001%, we will come to the same conclusion.

TABLE 8.4. p-Value and Association

p-Value	Gamma		
	<0.30	0.30 to 0.70	>0.70
<0.1	8	11	5
0.05	7	0	0
>0.10	8	0	0

One set of results is not more significant than another; it is only that the difference we uncovered was measurably more extreme in one set of samples than in another.

Note that, after examining the data, it is unethical to alter the significance level or to reinterpret a two-tailed test as if one had intended it to be one-tailed.

p-Values need not reflect the strength of a relationship. Duggan and Dean [1968] reviewed 45 articles that had appeared in sociology journals between 1955 and 1965 in which the chi-square statistic and the distribution had been employed in the analysis of 3×3 contingency tables, and they compared the resulting p-values with association as measured by Goodman and Kruskal's gamma. Table 8.4 summarizes their findings.

p-Values derived from tables are often crude approximations, particularly for small samples and tests based on a specific distribution. They and the stated significance level of our test may well be in error.

The vast majority of p-values produced by parametric tests based on the normal distribution are approximations. If the data are almost normal, the associated p-values will be almost correct. As noted in Chapter 6, the stated significance values for Student's t are very close to exact. Of course, a stated p-value of 4.9% might really prove to be 5.1% in practice. The significance values associated with the F-statistic can be completely inaccurate for nonnormal data (1% rather than 10%). And the p-values derived from the chi-square distribution for use with contingency tables also can be off by an order of magnitude.

The good news is that there exists a class of tests, the permutation tests described in Chapter 6, for which the significance levels are exact if the observations are i.i.d. under the null hypothesis or their labels are otherwise exchangeable.

Regardless of which test one uses, it is the height of foolishness to report p-values with excessive precision. For example, 0.06 and 0.052 are both acceptable, but 0.05312 suggests that you've let your software do the thinking for you.

CONFIDENCE INTERVALS

If p-values are misleading, what are we to use in their place? Jones [1955, p. 407] was among the first to suggest that "an investigator would be misled less frequently and would be more likely to obtain the information he seeks were he to formulate his experimental problems in terms of the estimation of population parameters, with the establishment of confidence intervals about the estimated values, rather than in terms of a null hypothesis against all possible alternatives." See also Gardner and Altman [1996], Matthews and Altman [1996], Feinstein [1998], and Poole [2001].

Confidence intervals can be derived from the rejection regions of our hypothesis tests, whether the latter are based on parametric or nonparametric methods. Suppose $A(\theta')$ is a $1 - \alpha$ level acceptance region for testing the hypothesis $\theta = \theta'$; that is, we accept the hypothesis if our test statistic T belongs to the acceptance region $A(\theta')$ and reject it otherwise. Let $S(X)$ consist of all the parameter values θ^* for which $T[X]$ belongs to the acceptance region $A(\theta^*)$. Then $S(X)$ is an $1 - \alpha$ level confidence interval for θ based on the set of observations $X = \{x_1, x_2, \ldots, x_n\}$.

The probability that $S(X)$ includes θ_o when $\theta = \theta_o$ is equal to $\Pr\{T[X] \in A(\theta_o)$ when $\theta = \theta_o\} \geq 1 - \alpha$.

As our confidence $1 - \alpha$ increases, from 90% to 95%, for example, the width of the resulting confidence interval increases. Thus, a 95% confidence interval is wider than a 90% confidence interval.

By the same process, the rejection regions of our hypothesis tests can be derived from confidence intervals. Suppose our hypothesis is that the odds ratio for a 2×2 contingency table is 1. Then we would accept this null hypothesis if and only if our confidence interval for the odds ratio includes the value 1.

A common error is to misinterpret the confidence interval as a statement about the unknown parameter. It is not true that the probability that a parameter is included in a 95% confidence interval is 95%. What is true is that if we derive a large number of 95% confidence intervals, we can expect the true value of the parameter to be included in the computed intervals 95% of the time. (That is, the true values will be included *if* the assumptions on which the tests and confidence intervals are based are satisfied 100% of the time.) Like the p-value, the upper and lower confidence limits of a particular confidence interval are random variables, for they depend upon the sample that is drawn.

IMPORTANT TERMS

Acceptance region, $A(\theta_o)$. A set of values of the statistic $T[X]$ for which we would accept the hypothesis H: $\theta = \theta_o$. Its complement is called the rejection region.

Confidence region, $S(X)$. Also referred to as a confidence interval (for a single parameter) or a confidence ellipse (for multiple parameters). A set of values of the parameter θ for which, given the set of observations $X = \{x_1, x_2, \ldots, x_n\}$ and the statistic $T[X]$, we would accept the corresponding hypothesis.

Confidence intervals can be used both to evaluate and report on the precision of estimates (see Chapter 5) and the significance of hypothesis tests (see Chapter 6). The probability that the interval covers the true value of the parameter of interest and the method used to derive the interval must also be reported.

In interpreting a confidence interval based on a test of significance, it is essential to realize that the center of the interval is no more likely than any other value, and that the confidence to be placed in the interval is no greater than the confidence we have in the experimental design and statistical test it is based upon. (As always, GIGO.)

Multiple Tests

Whether we report p-values or confidence intervals, we need to correct for multiple tests, as described in Chapter 6. The correction should be based on the number of tests we *perform*, which in most cases will be larger than the number on which we report. See Westfall and Young [1993], Aickin and Gensler [1996], and Hsu [1996] for a discussion of some of the methods that can be employed to obtain more accurate p-values.

Analysis of Variance

"An ANOVA table that contains only F-values is almost useless." Says Yoccoz [1991] who recommends that analysis of variance tables include estimates of standard errors, means, and differences of means, along with confidence intervals.

Don't ignore significant interactions. The guidelines on reporting the results of a multifactor analysis are clear-cut and too often ignored. If the interaction between A and B is significant, then the main effects of A should be calculated and reported separately for several levels of the factor B.

Or, to expand on the quote from George Dyke with which we opened this chapter, "Don't just cut out the appropriate part of the computer output and paste it onto the draft of the paper, but read it through and conduct the additional calculations suggested by the original analysis."

RECOGNIZING AND REPORTING BIASES

Very few studies can avoid bias at some point in sample selection, study conduct, and results interpretation. We focus on the wrong endpoints; participants and coinvestigators see through our blinding schemes; the effects of neglected and unobserved confounding factors overwhelm and outweigh the effects of our variables of interest. With careful and prolonged planning, we may reduce or eliminate many potential sources of bias, but seldom will we be able to eliminate all of them. Accept bias as inevitable and then endeavor to recognize and report all exceptions that do slip through the cracks.

Most biases occur during data collection, often as a result of taking observations from an unrepresentative subset of the population rather than from the population as a whole. The example of the erroneous forecast of Dewey's win over Truman in the 1948 presidential election was cited in Chapter 3. In Chapter 6, we considered a study that was flawed because of a failure to include planes that did *not* return from combat.

When analyzing extended time series in seismological and neurological investigations, investigators typically select specific cuts (a set of consecutive observations in time) for detailed analysis rather than trying to examine all the data (a near impossibility). Not surprisingly, such cuts usually possess one or more intriguing features not to be found in run-of-the-mill samples. Too often, theories evolve from these very biased selections. We expand on this point in Chapter 10 in discussing the limitations on the range over which a model may be applied.

Limitations in the measuring instrument such as censoring at either end of the scale can result in biased estimates. Current methods of estimating cloud optical depth from satellite measurements produce biased results that depend strongly on satellite viewing geometry. In this and similar cases in the physical sciences, absent the appropriate nomograms and conversion tables, interpretation is impossible.

Over- and underreporting plague meta-analysis (discussed in Chapter 7). Positive results are reported for publication; negative findings are suppressed or ignored. Medical records are known to underemphasize conditions such as arthritis, for which there is no immediately available treatment, while overemphasizing the disease of the day. [See, e.g., Callaham et al., 1998.]

Collaboration between the statistician and the domain expert is essential if all sources of bias are to be detected and corrected for, as many biases are specific to a given application area. In the measurement of price indices, for example, the three principal sources are substitution bias, quality change bias, and new product bias.[6]

Two distinct statistical bias effects arise with astronomical distance indicators (DI's), depending on the method used.[7]

> In one approach, the redshifts of objects whose DI-inferred distances are within a narrow range of some value d are averaged. Subtracting d from the resulting mean redshift yields a peculiar velocity estimate; dividing the mean redshift by d gives an estimate of the parameter of interest. These estimates will be biased because the distance estimate d itself is biased and is not the mean true distance of the objects in question.

> This effect is called homogeneous Malmquist bias. It tells us that, typically, objects lie further away than their DI-inferred distances. The physical cause is [that] more objects "scatter in" from larger true distances (where there is more volume) than "scatter out" from smaller ones.

> A second sort of bias comes into play because some galaxies are too faint or small to be in the sample; in effect, the large-distance tail of $P(d|r)$ is cut off. It follows that the typical inferred distances are smaller than those expected at a given true distance r. As a result, the peculiar velocity model that allows true distance to be estimated as a function of redshift is tricked into returning shorter distances. This bias goes in the same sense as Malmquist bias, but is fundamentally different. [It results not from volume/density effects, but from the same sort of sample selection effects that were discussed earlier in this section.]

---

[6] Otmar Issing in a speech at the Centre for Economic Policy Research/European Central Bank (CEPR/ECB) workshop on issues in the measurement of price indices, Frankfurt am Main, 16 November 2001.
[7] These next paragraphs are taken with minor changes from Willick [1999, Section 9].

Selection bias can be minimized by working in the "inverse direction." Rather than trying to predict absolute magnitude (Y) given a value of the velocity width parameter (X), one fits a line by regressing the widths X on the magnitudes Y.

Finally, bias can result from grouping or averaging data. Bias if group randomized trials are analyzed without correcting for cluster effects was reported by Feng et al. [1996]; see Chapter 6. The use of averaged rather than end-of-period data in financial research results in biased estimates of the variance, covariance, and autocorrelation of the first as well as higher-order changes. Such biases can be both time varying and persistent [Wilson, Jones, and Lundstrum, 2001].

REPORTING POWER

Statisticians are routinely forced to guess at the values of population parameters to make the power calculations needed to determine sample size. It's tempting, once the data are in hand, to redo these same power calculations. Do and don't.

Do repeat the calculations using the same effect size and variance estimate used originally while correcting for a reduced sample size due to missing data. On the other hand, post hoc calculations making use of parameter estimates provided by the data invariably inflate the actual power of the test [Zumbo and Hubley, 1998].

DRAWING CONCLUSIONS

Found data (nonrandom samples) can be very useful in suggesting models and hypotheses for further exploration. But without a randomized study, formal inferential statistical analyses are not supported [Greenland, 1990; Rothman, 1990]. The concepts of significance level, power, p-value, and confidence interval apply only to data that have arisen from carefully designed and executed experiments and surveys.

A vast literature has grown up around the unease researchers feel in placing too much reliance on p-values. Examples include Selvin [1957], Yoccoz [1991], McBride, Loftis, and Adkins [1993], Nester [1996], Suter [1996], Feinstein [1998], Badrick and Flatman [1999], Johnson [1999], Jones and Tukey [2000], and Parkhurst [2001].

The vast majority of such cautions are unnecessary, provided that we treat p-values as merely one part of the evidence to be used in decision making. They need to be viewed and interpreted in the light of all the surrounding evidence, past and present. No computer should be allowed to make decisions for you.

A failure to reject may result from any of the following:

1. A Type II error.
2. Insensitive or inappropriate measurements.
3. Additional variables being confounded with the variable of interest.
4. Too small a sample size.

This last is another reason why the power of your tests should always be reported after correcting for missing data.

A difference that is statistically significant may be of no practical interest. Take a large enough sample and we will always reject the null hypothesis; take too small a sample and we will never reject. To say nothing of "significant" results which arise solely because their authors chose to test a "null" hypothesis rather than one of practical interest. (See Chapter 5.)

Many researchers would argue that there are always three regions to which a statistic may be assigned: acceptance, rejection, and indifference. When a statistic falls in the last intermediate region, it may suggest a need for additional experiments. The p-value is only one brick in the wall; all our other knowledge must and should be taken into consideration [Horwitz et al., 1998].

Finally, few journals publish negative findings, so avoid concluding that "most studies show."

SUMMARY

- Provide details of power and sample size calculations.
- Describe treatment allocation.
- Detail exceptions, including withdrawals and other sources of missing data.
- Use meaningful measures of dispersion.
- Use confidence intervals in preference to p-values.
- Report sources of bias.
- Formal statistical inference is appropriate only for randomized studies and predetermined hypotheses.

TO LEARN MORE

The book by Lang and Secic [1997] is must reading; reporting criteria for meta-analyses are given on pages 177ff. See Tufte [1983] on the issue of tables versus graphs. For more on the geometric versus the arithmetic mean, see Parkhurst [1998]. For more on reporting requirements, see Bailar and Mosteller [1988], Grant [1989], Begg et al. [1996] the International Committee of Medical Journal Editors [1997], and Altman et al. [2001; the revised CONSORT statement].

Mosteller [1979] and Anderson and Hauck [1986] warn against the failure to submit reports of negative or inconclusive studies and the failure of journal editors to accept them. To address this issue, the *Journal of Negative Results in Biomedicine* has been launched at http://www.jnrbm.com/start.asp.

On the proper role of p-values, see Neyman [1977], Cox [1977], www.coe.tamu.edu/~bthompson; www.indiana.edu/~stigsts; and www.nprc.ucgs.gov/perm/hypotest, Poole [1987, 2001].

To learn more about decision theory and regions of indifference, see Duggan and Dean [1968] and Hunter and Schmidt [1997].

REQUESTED MANUSCRIPT FORMATS

For Submission to *Academic Emergency Medicine* As posted at http://www.aemj.org/misc/reqmanfor.shtml

Study Protocol Describe the method of patient enrollment (i.e., consecutive, convenience, random, population sampling); discuss any consent process; note any interventions used; describe any blinding or randomization regarding treatments, purpose of the study, or data collection; discuss if and how standard treatment was administered (describe such standard treatment separate from interventions used specifically for study purposes), and placebo specifics (how prepared, delivered) and the reasoning for such (especially if the study is an analgesia trial).

Measurements Discuss the data collection. Clarify who collected the data. Describe any special data collection techniques or instruments. Provide [the] manufacturer's name and address along with [the] brand name and model number for equipment used in the study. Denote what instructions or training the data collectors were given.

Data Analysis Summarize how the major outcome variables were analyzed (clearly define outcome measures). If multiple definitions must be provided, include a separate subheading for definitions. Note which outcomes were analyzed with which statistical tests. Clearly define any criterion standards (do not use the phrase "gold' standard"). Note any important subgroup analyses and whether these were planned before data collection or arose after initial data evaluation. Denote any descriptive statistics used. Provide 95% confidence intervals for estimates of test performance where possible; they should be described as 95% CI = X to X. Discuss sample size estimates. Note significance levels used.

9

INTERPRETING REPORTS

Smoking is one of the leading causes of statistics.

—Knebel

The previous chapter was aimed at practitioners who must prepare reports. This chapter is aimed at those who must read them.

WITH A GRAIN OF SALT

Critics may complain that we advocate interpreting reports not merely with a grain of salt but with an entire shaker; so be it. Internal as well as published reports are the basis of our thought processes, not just our own publications. Neither society nor we can afford to be led down false pathways.

We are often asked to testify in court or to submit a pretrial declaration in which we comment on published reports. Sad to say, too often the reports our clients hand us lack even the most basic information such as sample sizes, measures of variability, and descriptions of statistical methods—and this despite their having appeared in refereed publications!

The Authors

Begin your appraisal with the authors' affiliations: Who conducted this study? What is their personal history in conducting other studies? What personal interest might they have in the outcome?

Who funded the study? What is the funding agency's history regarding other studies, and what is their interest in the outcome?

Common Errors in Statistics (and How to Avoid Them), Third Edition. Edited by P. I. Good and J. W. Hardin
Copyright © 2009 John Wiley & Sons, Inc.

The Samples

What population(s) was/were sampled from? Were these the same populations to which the report's conclusions were applied?

For example, studies of the possibilities of whiplash resulting from low-speed rear-end collisions would be relevant to specific court cases only if the subjects of the studies were of the same age and physical condition and had the same history of prior injuries as the subjects in the court cases, and if the speeds of impact, and the masses and protective ability of the vehicles involved, were the same in both the studies and the court cases.

How large was the sample? This most basic piece of information is missing from the report by Okano et al. [1993]. Was the sample random? Stratified? Clustered? What was the survey unit? Was the sample representative? Can you verify this from the information provided in the report?

For example, when several groups are to be compared, baseline information for each group should be provided. A careful reading of Fujita et al. [2000] reveals that the baseline values (age and bone density) of the various groups were quite different, casting doubt on the reported findings.

How was the sample size determined? Was the anticipated power stated explicitly? Without knowledge of the sample size and the anticipated power of the test, we will be unable to determine what interpretation, if any, ought be given a failure to detect a statistically significant effect.

What method of allocation to subgroups was used? What type of blinding was employed, if any? How was blinding verified?

With regard to surveys, what measures were taken to ensure that the responses of different individuals were independent of one another? What measures were taken to detect lies and careless responses? Were nonresponders contacted and interviewed?

Are the authors attempting to compare or combine samples that have been collected subject to different restrictions and by different methods?

Aggregating Data

Just as the devil often quotes scripture for his (or her) own purposes, politicians and government agencies are fond of using statistics to mislead. One of the most common techniques is to combine data from disparate sources. Four of the six errors reported by Wise [2005] in the presentation of farm statistics arise in this fashion. (The other two come from employing arithmetic means rather than medians in characterizing highly skewed income distributions.) These errors are:

1. Including "Rural Residence Farms," which represent two-thirds of all U.S. farms but do *not* farm for a living, in the totals for the farm sector. As Wise notes, "This leads to the misleading statement that a minority of farms get farm payments. A minority of *part-time* farmers get payments, but a significant majority of full-time commercial and family farmers receives farm payments".
2. Including income from nonfarming activities in farm income.
3. Attributing income to farmers that actually goes to landowners.

4. Mixing data from corporate farms with those of multimember collective entities such as Indian tribes and cooperatives.

Experimental Design

If a study is allegedly double-blind, how was the blinding accomplished?
Are all potential confounding factors listed and accounted for?

WHAT IS THE SOUND OF ONE HAND CLAPPING?

Gonzales et al. [2001] reported that Maca improved semen parameters in men. A dozen men were treated with Maca. But no matched untreated (control) subjects were studied during the same period. Readers and authors will never know whether the observed effects were due to a change in temperature, a rise in the Peruvian economy, or several dozen other physiological and psychological factors that might have been responsible for the change. (Our explanation for the reported results is that 12 men who normally would have their minds occupied by a score of day-to-day concerns spent far more time than usual thinking about sex and the tests to come. Hence the reported increase in semen production.)

The big question is not why this article was published in the absence of a control group, but why the human-uses committee at the Universadad Peruna Cayento Heredia in Lima permitted the experiments to go forth in the first place. The tests were invasive—"semen samples were collected by masturbation." A dozen men were placed at risk and subjected to tests, yet the final results were (predictably) without value.

Descriptive Statistics

Is all the necessary information present? Measures of dispersion (variation) should be included as well as measures of central tendency. Was the correct and appropriate measure used in each instance—mean (arithmetic or geometric) or median? Standard deviation or standard error or bootstrap confidence interval?

Are missing data accounted for? Does the frequency of missing data vary among treatment groups?

THE ANALYSIS

Tests

Authors must describe which test they used, report the effect size (the appropriate measure of the magnitude of the difference, usually the difference or ratio between groups; a confidence interval would be best), and give a measure of significance, usually a p-value, or a confidence interval for the difference.

Can you tell which tests were used? Were they one-sided or two-sided? Was this latter choice appropriate? Consider the examples we considered in Tables 6.1a, b and 6.2.

How many tests? In a study by Olsen [2003] of articles in *Infection and Immunity* the most common error was a failure to adjust or account for multiple comparisons. Remember, the probability is 64% that at least one test in 20 is likely to be significant at the 5% level by chance alone. Thus, it's always a good idea to check the methods section of an article to see how many variables were measured. See, O'Brien [1983], Saville [1990], Tukey [1991], Aickin and Gensler H [1996], and Hsu [1996], as well as the minority views of Rothman [1990] and Saville [2003].

Was the test appropriate for the experimental design? For example, was a matched-pairs *t*-test used when the subjects were not matched?

Note to journal editors: The raw data that formed the basis for a publication should eventually be available on a website for those readers who may want to run their own analyses.

Contingency Tables

Was an exact method used for their analysis or a chi-square approximation?

Were log-linear models used when the hypothesis of independence among diverse dimensions in the table did not apply?

Factor Analysis

Was factor analysis applied to data sets with too few cases in relation to the number of variables analyzed?

Was oblique rotation used to get a number of factors bigger or smaller than the number of factors obtained in the initial extraction by principal components as a way to show the validity of a questionnaire? An example provided by Godino, Batanero, and Gutiérrez-Jaimez [2001] is obtaining only one factor by principal components and using the oblique rotation to justify that there were two differentiated factors, even when the two factors were correlated and the variance explained by the second factor was very small.

Multivariate Analysis. One should always be suspicious of a multivariate analysis, both of the methodology and of the response variables employed. While Student's, *t* is very robust, even small deviations from normality make the *p*-values obtained from Hotelling's T^2 suspect. The inclusion of many irrelevant response variables in a recent multivariate permutation test of the effects of industry on the levels of pollutant off the coasts of Sicily resulted in values that were not statistically significant.

Correlation and Regression

Always look for confidence intervals about the line. If they aren't there, distrust the results unless you can get hold of the raw data and run the regressions and a bootstrap validation yourself (see Chapters 13 and 14).

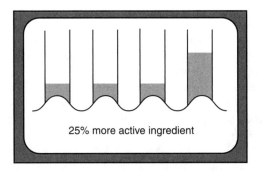

Figure 9.1. Missing baseline maker true comparisons impossible.

Graphics

Beware of missing baselines, as in Figure 9.1. Be wary of extra dimensions that inflate relative proportions. (See Chapter 10.) Distrust curves that extend beyond the plotted data. Check to see that charts include all data points, not just some of them.

The data for Figure 9.2 supplied by the California Department of Education are accurate. The title added by an Orange County newspaper is not. While enrollment in the Orange County public schools may have been steadily increasing in the last quarter of the twentieth century, clearly it has begun to level off and even to decline in the twenty-first.

Finally, as noted in the next chapter, starting the y-axis at 150,000 rather than 0 gives the misleading impression that the increase is much larger than it actually is.

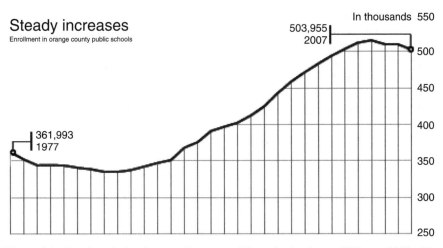

Figure 9.2. Enrollment in Orange County public schools from 1977 to 2007. The misleading title was added by an Orange County newspaper. *Source*: California Department of Education.

Conclusions

Our greatest fault (apart from those to which our wives have been kind enough to draw our attention) is to save time by relying on the abstract and/or the summary of a paper for our information rather than wade through the entire article. After all, some reviewer has already gone through it with a fine-tooth comb. Or has he? Most reviewers, though experts in their own disciplines, are seldom equally knowledgeable about statistics. It's up to us to do the job, to raise the questions that ought to have been asked before the article was published.

Has an attempt been made to extend the results beyond the populations that were studied? Have potential biases been described?

Were any of the tests and subgroup analyses performed after the data were examined, thereby rendering the associated p-values meaningless? And, again one need ask, were all potential confounding factors accounted for, either by blocking or by treatment as covariates. (See, for example, the discussion of Simpson's paradox in Chapter 11.)

Be wary of extrapolations, particularly in multifactor analyses. As the small print reads on a stock prospectus, past performance is no guarantee of future success.

Are non-significant results taken as proof of lack of effect? Are practical and statistical significance distinguished?

Finally, few journals publish negative findings, so avoid concluding that "most studies show."

THE COURTS EXAMINE THE SAMPLING UNIVERSE

The U.S. Equal Employment Opportunities Commission (EEOC) alleged that Eagle Iron Works assigned African-Americans to unpleasant work tasks because of their race and discharged African-Americans in greater numbers than Caucasians, again because of their race.[1] The EEOC was able to identify only 1200 out of 2000 past and present employees by race, though all 250 current employees could be identified. The Court rejected the contention that the 250 current employees were a representative sample of all 2000; it also rejected the EEOC's unsubstantiated contention that all unidentified former workers were Caucasian. "The lack of a satisfactory basis for such an opinion and the obvious willingness of the witness to attribute more authenticity to the statistics than they possessed, cast doubts upon the value of opinions."

The plaintiff's survey was rejected in Bristol Meyers v. FTC[2] as there was no follow up of the 80% of those who did not respond.

Amstar Corporation claimed that "Domino's Pizza" was too easily confused with its own use of the trademark "Domino" for sugar. The Appeals Court

---

[1] *Eagle Iron Works*, 424 F. Supp, 240 (S.D. Ia. 1946).
[2] 185 F.2d 258 (4th Cir. 1950).

found that the surveys both parties used to support their claims were substantially defective.

"In undertaking to demonstrate likelihood of confusion in a trademark infringement case by use of survey evidence, the appropriate universe should include a fair sampling of those purchasers most likely to partake of the alleged infringer's goods or service."[3]

Amstar conducted and offered in evidence a survey of heads of households in ten cities. But Domino's Pizza had no stores or restaurants in eight of these cities, and in the remaining two, their outlets had been open less than three months. Only women were interviewed by Amstar, and only those women who were at home during daylight hours, that is, grocery shoppers rather than the young and the single that compose the majority of pizza eaters. Similarly, the court rejected Domino's Pizza's own survey conducted in its pizza parlors. Neither plaintiff nor defendant had sampled from a sufficiently complete universe.

RATES AND PERCENTAGES

Consider the statement "Sixty percent of the children in New York City read below grade level!" Some would say we can't tell whether this percentage is of practical significance without some means of comparison. How does New York City compare with other cities of its size? What about racial makeup? What about other environmental factors compared with those of other similar cities?

In the United States in 1985, there were 2.1 million deaths from all causes, compared to 1.7 million in 1960. Does this mean that it was safer to live in the United States in the 1960s than in the 1980s? We don't know the answer, because we don't know the sizes of the U.S. population in 1960 and 1985.

If a product had a 10% market share in 1990 and a 15% share today, is this a 50% increase or a 5% increase? Not incidentally, note that market share may increase even when total sales decline.

How are we to compare rates? If a population consists of 12% African-Americans and a series of jury panels contain only 4%, the absolute disparity is 8% but the comparative disparity is 66%.

In *Davis v. City of Dallas*,[4] the court observed that a "7% difference between 97% and 90% ought not to be treated the same as a 7% difference between, for example 14% and 7%, since the latter figure indicates a much greater degree of disparity". Not so, for pass rates of 97% and 90% immediately imply failure rates of 3% and 10%.

The consensus among statisticians is that one should use the odds ratio for such comparisons defined as the percentage of successes divided by the percentage of failures. In the present example, one would compare $97\%/3\% = 32.3$ versus $90\%/10\% = 9$.

[3] Amstar Crop. v. Domino's Pizza, Inc., 205 U.S.P.Q 128 (N.D. Ga. 1979), rev'd, 615 F. 2d 252 (5th Cir. 1980).
[4] 487 F. Supp 389 (N.D. Tex 1980).

INTERPRETING COMPUTER PRINTOUTS

Many of our reports come to us directly from computer printouts. Even when we're the ones who've collected the data, these reports are often a mystery. One such report, generated by SAS PROC TTEST, is reproduced and annotated below. We hope that our annotations will inspire you to do the same with the reports your software provides you. (*Hint*: Read the manual.)

First, a confession: We've lopped off many of the decimal places that were present in the original report. They were redundant, as the original observations had only two decimal places. In fact, the fourth decimal place is still redundant.

Second, we turn to the foot of the report, where we learn that a highly significant difference was detected between the dispersions (variances) of the two treatment groups. We'll need to conduct a further investigation to uncover why this is true.

Confining ourselves to the report in hand, unequal variances mean that we need to use the Satterthwaite's degrees of freedom adjustment for the *t*-test for which $\Pr > |t| = 0.96$, that is, the values of RIG for the New and Standard treatment groups are not significantly different from a statistical point of view.

Lastly, the seventh line of the report tells us that the difference in the means of the two groups is somewhere in the interval $(-0.05, +0.05)$. (The report does not specify what the confidence level of this confidence interval is, and we need to refer to the SAS manual to determine that it is 95%.)

```
                    The TTEST Procedure
                       Statistics

              Lower CL        Upper CL  Lower CL       Upper CL
Var'le  treat  N  Mean   Mean  Mean   Std Dev  Std Dev Std Dev  Std Err
RIG     New   121 0.5527 0.5993 0.6459 0.2299   0.2589  0.2964   0.0235
RIG     Stand 127 0.5721 0.598  0.6238 0.1312   0.1474  0.1681   0.0131
RIG     Diff (1-2)-0.051 0.0013 0.0537 0.1924   0.2093  0.2296   0.0266
```

```
                       T-Tests
        Variable   Method              Variances   DF   t Value   Pr>|t|
        RIG        Pooled              Equal       246  0.05      0.9608
        RIG        Satterthwaite       Unequal     188  0.05      0.9613
```

```
                  Equality of Variances
        Variable   Method      Num DF   Den DF   F Value   Pr>F
        RIG        Folded F    120      126      3.09      <.0001
```

TO LEARN MORE

Godino, Batanero, and Gutiérrez-Jaimez [2001] report on the errors found in the use of statistics in a sample of mathematics education doctoral theses in Spain.

CRITICIZING REPORTS

Commenting on an article by Rice and Griffin [2004], David Hershey (http://www.fortunecity.com/greenfield/clearstreets/84/hornworm.htm) cites the following flaws:

1. Using the arithmetic average (linear interpolation) of two values that do not fall on a straight line.
2. Plotting curves without plotting the corresponding confidence intervals.
3. Failure to match treatment groups based on baseline data. As a result, such factors as the weight of the subject were confounded with treatment.
4. No explanation provided for the missing data (occasioned by the deaths of the experimental organisms).
5. No breakdown of missing data by treatment.
6. Too many significant figures in tables and equations.
7. Extrapolation leading to a physiologically impossible end point.
8. Concluding that detecting a significant difference provided confirmation of the validity of the experimental method.

10

GRAPHICS

KISS—Keep it simple but scientific.

—Parzen

Getting information from a table is like extracting sunbeams from a cucumber.

—Farquhar and Farquhar

How many dimensions do you really need to illustrate? Do you need to illustrate repeated information for several groups? Is a graphical illustration a better vehicle than a table for communicating information to the reader? How do you select from a list of competing choices? How do you know whether the graphic is effectively communicating the desired information?

Graphics should emphasize and highlight salient features of the underlying data and should coherently summarize large quantities of information. While graphics provide a break from dense prose, authors must not forget that these illustrations should be scientifically informative rather than decorative. In this chapter, we outline mistakes in selection, creation, and execution of graphics and then discuss improvements.

Graphical illustrations should be simple and pleasing to the eye, but motivation for their inclusion must remain scientific. In other words, we avoid having too many graphical features that are purely decorative while keeping a critical eye open for opportunities to enhance the scientific implications for the reader. Good graphical designs utilize a large proportion of the ink to communicate scientific information in the overall display.

Common Errors in Statistics (and How to Avoid Them), Third Edition. Edited by P. I. Good and J. W. Hardin
Copyright © 2009 John Wiley & Sons, Inc.

THE SOCCER DATA

When his children were young, Dr. Hardin coached youth soccer teams (players of age five) and recorded the total number of goals scored for the top five teams during the eight-game spring 2001 season in College Station, Texas. The total number of goals scored per team was 16 (Team 1), 22 (Team 2), 14 (Team 3), 11 (Team 4), and 18 (Team 5). There are many ways we can describe these outcomes to the reader. Above, we simply communicated them in words.

A more effective presentation would be to write "The total number of goals scored by Teams 1 through 5 was 16, 22, 14, 11, and 18, respectively." The College Station Soccer Club assigned the official team names Team 1, Team 2, etc.[1] Improving on this textual presentation, we could also write "With the team number as the subscript, the total number of goals was 22_2, 18_5, 16_1, 14_3, and 11_4." This presentation improves communication by ordering the outcomes. With these particular data, the reader will naturally want to know the order.

FIVE RULES FOR AVOIDING BAD GRAPHICS

There are a number of choices in presenting the soccer outcomes in graphical form. Many are poor choices; they hide information, make it difficult to discern actual values, or use the space within the graphic inefficiently. Open almost any newspaper and you will see a bar chart similar to Figure 10.1a which illustrates the soccer data. In this section, we provide five important rules for generating effective graphics. Subsequent sections will augment this list with specific examples.

Figure 10.1a includes a third dimension, a depth dimension that does not correspond to any information in the data. The resulting figure obscures the outcomes. Does Figure 10.1a indicate that Team 3 scored 14 goals, or does it appear that the team scored 13 goals? The reader must focus on the top back corner of the three-dimensional rectangle since that part of the bar is (almost) at the same level as the grid lines on the plot; actually, the reader must first focus on the floor of the plot to initially differentiate the vertical distance of the back right corner of the rectangular bar from the corresponding grid line at the back (these are at the same height). The reader must then mentally transfer this difference to the top of the rectangular bars to accurately infer the correct value.

To highlight the confusing effect caused by the false third dimension, look at Figure 10.1b, where we provided additional grid lines. This plot illustrates the previously described technique for how to infer values from this type of graphic. The reality is that most readers focus on the front face of the rectangle and subsequently misinterpret values in this data representation.

---

[1] These labels show the remarkable lack of imagination that we encounter in many data collection efforts. To be fair, the children had their own informal names, such as Fireballs, but not all of these names were available at data collection time.

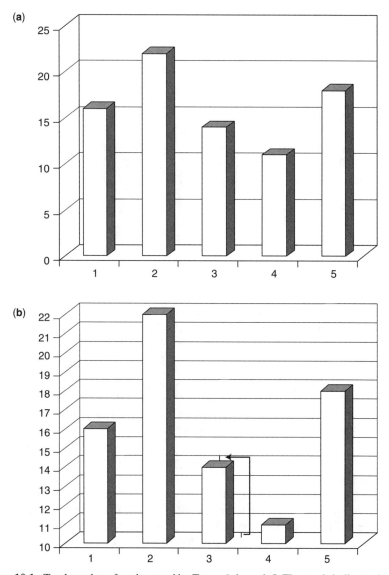

Figure 10.1. Total number of goals scored by Teams 1 through 5. The *x*-axis indicates the team number, and the *y*-axis indicates the number of goals scored by the respective team.

(a) *Problem*: The false third dimension makes it difficult to discern values. The number of goals for Team 3 appears to be 13 rather than the correct value of 14.

(b) *Problem*: The false third dimension makes it difficult to discern values.

Solution: Compute the vertical distance from the back-right bottom corner of a bar to the first vertical value. Transfer this value to the top of the back face of a bar. This height may then be compared to the added gridlines so that the correct value (14) may be inferred from the graphic.

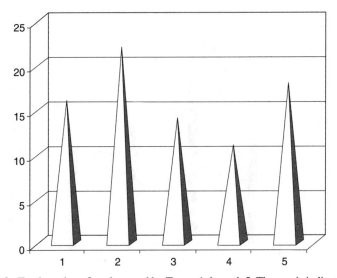

Figure 10.2. Total number of goals scored by Teams 1 through 5. The *x*-axis indicates the team number, and the *y*-axis indicates the number of goals scored by the respective team.
Problem: The false third dimension makes it difficult to discern the values in the plot. Since the back face is the most important one for interpreting the values, the fact that the decorative object comes to a point makes it impossible to read values correctly from the plot.

Figure 10.2 also includes a false third dimension. As in the previous example, the resulting illustration makes it difficult to discern the actual values presented. This problem is further complicated by the fact that the depth dimension has been eliminated at the top of the three-dimensional pyramids so that it's nearly impossible to correctly ascertain the plotted values. Focus on the result of Team 4, compare it to the illustration in Figure 10.1a, and judge whether you think that the plots are using the same data (they are).

Other types of plots that confuse the reader (and writer) with false third dimensions include point plots with shadows and line plots where the data are connected with a three-dimensional line or ribbon. The only sure way to fix the problems in Figure 10.2 is to include the values atop each pyramid as a textual element or to include a tabular legend with the values.[2]

The point of these graphics is to avoid illustrations that utilize more dimensions than exist in the data. Clearly, a better presentation would indicate only two dimensions, one identifying the teams and the other the number of goals scored.

Rule 1: Don't produce graphics illustrating more dimensions than exist in the information to be illustrated.

---

[2]If we include all of the values as text (as labels or in a tabular legend), the graph should illustrate more than just the labeled values.

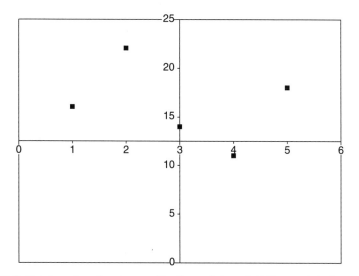

Figure 10.3. Total number of goals scored by Teams 1 through 5. The *x*-axis indicates the team number, and the *y*-axis indicates the number of goals scored by the respective team.
Problem: Placing the axes inside the plotting area effectively occludes data information. This violates the simplicity goal of graphics; the reader should be able to see easily all of the numeric labels in the axes and plot region.

Figure 10.3 is an improvement over three-dimensional displays. It is easier to discern the outcomes for the teams, but the axis label obscures the outcome of Team 4. Axes should be moved outside of the plotting area with enough labels so that the reader can quickly scan the illustration and identify the values.

> *Rule 2: Don't superimpose labeling information on the graphical elements of interest. Labels add information to the plot, but they should be placed in (otherwise) unused portions of the plotting region.*

Figure 10.4 is a much better display of the information of interest. However, this graphic suffers from too much empty space. Beginning the vertical axis at zero means that about 40% of the plotting region is empty. Unless there is a scientifically compelling reason to include a specific baseline in the graph, the presentation should be limited to the range of the information at hand. You can ignore this rule if you want to include zero as the baseline to admit a relative comparison of the values as well as an absolute comparison. Note that the symbol for Team 2 is twice as high as the symbol for Team 4 in Figure 10.4, but in Figure 10.5 this is no longer true since we eliminate the zero range of the data. There are several cases in which axis range can exceed the information at hand, and we will illustrate those in a presentation.

> *Rule 3: Don't allow the range of the axis labels to significantly decrease the area devoted to data presentation. Choose limits wisely and do not accept*

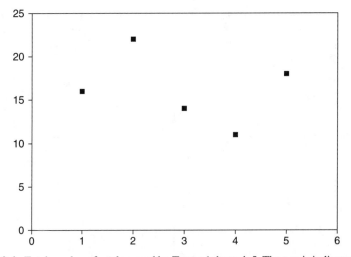

Figure 10.4. Total number of goals scored by Teams 1 through 5. The *x*-axis indicates the team number, and the *y*-axis indicates the number of goals scored by the respective team.
Problem: By allowing the *y*-axis to range from zero, the presentation reduces the proportion of the plotting area in which we are interested. Less than half of the vertical area of the plotting region is used to communicate data.

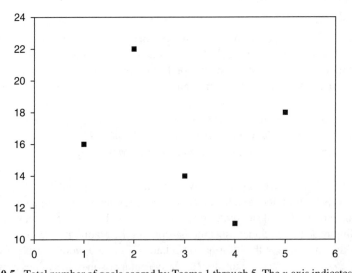

Figure 10.5. Total number of goals scored by Teams 1 through 5. The *x*-axis indicates the team number, and the *y*-axis indicates the number of goals scored by the respective team.
Problem: This graph correctly scales the *y*-axis, but still uses a categorical variable denoting the team on the *x*-axis. Labels 0 and 6 do not correspond to a team number, and the presentation appears as if the *x*-axis is a continuous range of values when in fact it is merely a collection of labels. While this is a reasonable approach to communicating the desired information, we can improve on this presentation by changing the numeric labels on the *x*-axis to string labels corresponding to the team names.

default values for the axes that are far outside of the range of data unless
relative as well as absolute comparisons should be made by the reader.

Figure 10.5 eliminates the extra space included in Figure 10.4 where the vertical
axis is allowed to match the range of the outcomes more closely. The presentation
is good, but it could be better. The data of interest in this case involve a continuous
and a categorical variable. This presentation treats the categorical variable as numeric
for the purposes of organizing the display, but this is not necessary.

Rule 4: Carefully consider the nature of the information underlying the axes.
Numeric axis labels imply a continuous range of values that can be con-
fusing when the labels actually represent discrete values of an underlying
categorical variable.

Figures 10.5 and 10.6 are further improvements of the presentation. The graph
region, that is, the area of the illustration devoted to the data, is illustrated with axes
that match the range of the data more closely. Figure 10.6 connects the point information with a line that may help visualize the difference between the values, but it
also indicates a nonexistent relationship; the horizontal axis is discrete rather than

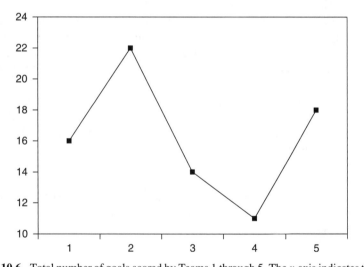

Figure 10.6. Total number of goals scored by Teams 1 through 5. The *x*-axis indicates the team
number, and the *y*-axis indicates the number of goals scored by the respective team.
Problem: The inclusion of a polyline connecting the five outcomes helps the reader to visualize
changes in scores. However, the categorical values are not ordinal, and the polyline indicates an
interpolation of values that does not exist across the categorical variable denoting the team
number. In other words, there is no reason that Team 5 is to the right of Team 3 other than
the fact that we ordered them that way, and there is no Team 3.5, as the presentation seems to
suggest.

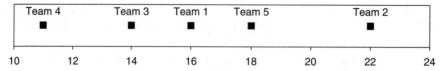

Figure 10.7. Total number of goals scored by Teams 1 through 5. The *x*-axis indicates with a square the number of goals scored by the respective team. The associated team name is indicated above the square. Labeling the outcomes addresses the science of the KISS specification given at the beginning of the chapter.

continuous. Even though these presentations vastly improve the illustration of the desired information, they are still two-dimensional. In fact, our data are not really two-dimensional, and the final illustration reflects the true nature of the information more accurately.

> *Rule 5: Do not connect discrete points unless there is either a scientific meaning to the implied interpolation or a collection of profiles for group-level outcomes.*

Rules 4 and 5 are aimed at the practice of substituting numbers for labels and then treating those numeric labels as if they were in fact numeric. Had we included the word "Team" in front of the labels, there would be no confusion about the nature of the labels. Even when nominative labels are used on an axis, we must consider the meaning of values between the labels. If the labels are truly discrete, data outcomes should not be connected or they may be misinterpreted as implying a continuous rather than a discrete collection of values.

Figure 10.7 is an excellent, and spatially economical, illustration of the soccer data. There are no false dimensions, the range of the graphic is close to the range of the data, there is no difficulty interpreting the values indicated by the plotting symbols, and the legend fully explains the material.

Table 10.1 succinctly presents the relevant information in tabular form. Tables and figures have the advantage over in-text descriptions that the information is more easily found while scanning through the document. If the information is summary in nature, we should make it easy to find and place it in a figure or table. If the information is ancillary to the discussion, it can be left in text.

TABLE 10.1. Total Numbers of Goals Scored by Teams 1 to 5 Ordered by Lowest Total to Highest Total

Team 4	Team 3	Team 1	Team 5	Team 2
11	14	16	18	22

Note: These totals are for the spring 2001 season. The organization of the table correctly sorts on the numeric variable. That the team labels are not sorted is far less important since these labels are merely nominal; without labeling with integers, the team names would have no natural order.

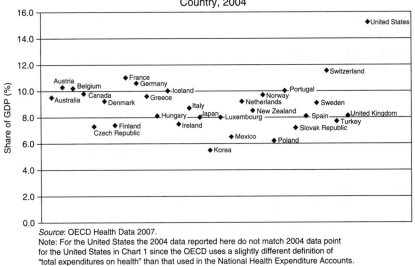

Figure 10.8. This chart prepared by the U.S. Office of the Actuary of the Department of Health and Human Services violates virtually all the rules.

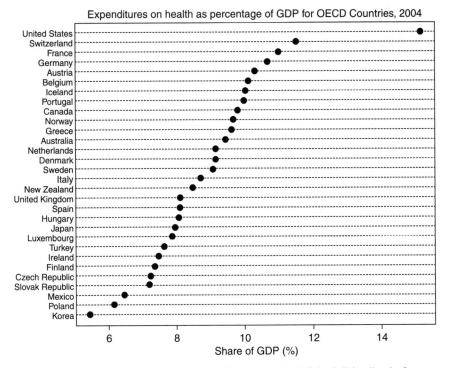

Figure 10.9. This chart, generated from an R program by Michael Friendly, is far more effective.

Figure 10.8, from a report by the Office of the Actuary of the Department of Health and Human Services, violates almost all the previous rules. Figure 10.9, a dotchart prepared by Michael Friendly in the R program, with country names on the vertical axis and the percentage of GDP spent on health on the horizontal axis, is far more effective, because it moves the country names outside the plot frame, makes the country dimension explicit, and sorts on the numeric values rather than the labels.

Choosing Between Tabular and Graphical Presentations

In choosing between tabular and graphical presentations, there are two issues to consider: the size (density) of the resulting graphic and the scale of the information. If the number of rows for a tabular presentation requires more than one page, the graphical presentation is usually preferred. Conversely, if the amount of information is small, the table is preferred. If the scale of the information makes it difficult to discern otherwise significant differences, a graphical presentation is better.

Study No.	Study Name	No. Events / No. Entered Treatment	Control	O–E	Variance	Odds Ratio ‡ (Treatment : Control)	Odds Redn. (±S.D.)
64B	Oslo	253/583	236/552	3.3	89.0		
69A	Heidelberg XRT	63/84	40/58	4.7	17.9		
70A₁	Manchester RBS₁	169/355	183/359	−7.2	67.5		
70B	Kings/Cambridge	676/1376	677/1424	14.6	269.2		
71B	StockholmA	119/323	132/321	−6.7	53.4		
71C₁	NSABP B-04*	145/352	170/365	−5.4	66.5		
72A	WSSA Glasgow	48/94	67/123	−3.9	21.3		
73A	Wessex	27/71	40/75	−6.9	13.9		
73C	Mayo Clinic	56/121	53/120	−0.9	22.3		
74B	Edinburgh I	50/173	50/175	1.8	22.2		
75K	Piedmont OA	71/145	56/136	4.9	27.3		
76A₂₋₃	SECSG 1	39/127	52/129	−4.0	19.4		
76C	Glasgow	44/112	42/102	0.0	18.3		
78A	S Swedish BCG	69/386	78/382	−6.3	34.6		
78D₄	Scottish D	13/74	5/46	3.8	4.3		
78G	CCABC Canada	25/154	24/137	−2.1	11.0		
82B₁	Danish BCG 82b	16/224	15/212	0.7	7.3		
82C	Danish BCG 82c	19/232	18/233	0.5	8.8		
84A₂	BMFT 03 Germany	0/22	0/23				
	Total	1902/ 4961	1938/ 4972	−9.0	774.2		1% ±4

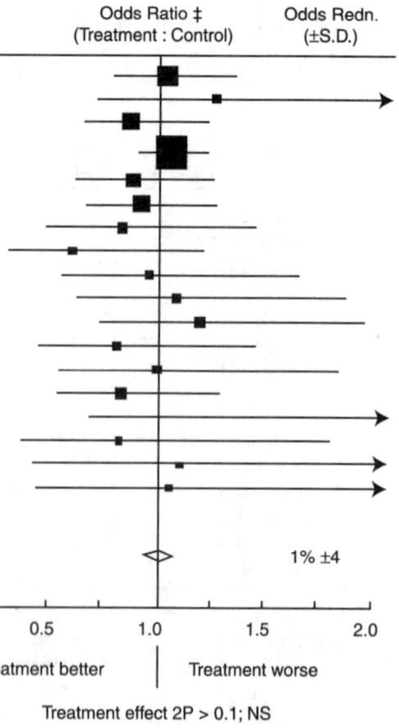

0.0   0.5   1.0   1.5   2.0

Test for heterogeneity: χ^2_{17} = 15.1: 2P > 0.1: NS   Treatment better | Treatment worse

Treatment effect 2P > 0.1; NS

* Published results (3), since individual patient data not available.

† Data from about 10 randomized radiotherapy trials that began before 1.1.1985 were not available in1985 and are not included here. Data from one large trial (Manchester Christie 49B) have been excluded because of non-standard randomization.

‡ 95% confidence intervals for overview and 99% individual trials.

Figure 10.10. An Overview of 19 clinical trials. Mortality in all available unconfounded randomized post-masectomy radiotherapy trials. Reproduced with permission from Oxford Univerisity Press from Table 3M of Early Breast Cancer Trialists' Collaborative Group [1990].

KISS

A picture may be worth 1000 words; but it shouldn't take 1000 words to explain your pictures.

Figure 10.10 summarizes the results of 19 clinical studies on the effects of radiotherapy on the survival of postmastectomy patients. The figure is a hybrid presentation in which tabular information is combined with a graphic.

But the figure is too ambitious and raises more questions than it answers. The axes for the odds ratio are asymmetric without explanation; that is, they are symmetric about the one for absolute values but not about the one for ratio values. The sizes (spacing) of the graphic also change without explanation. Because the information doesn't quite fit into the framework in which it is forced, three footnotes are required: one for a row, one for a column, and one for the overall title. Every possible way the reader might view the graphic proves to be a special case.

Curb your enthusiasm: *Keep it simple.*

Knowin' all the words in the dictionary ain't gonna help if you got nuttin' to say.

—Blind Lemon Jefferson

ONE RULE FOR CORRECT USAGE OF THREE-DIMENSIONAL GRAPHICS

As illustrated in the previous section, the introduction of superfluous dimensions in graphics should be avoided. The prevalence of turnkey solutions in software that implement these decorative presentations is alarming. At one time, these graphics were limited to business-oriented software and presentations, but this is no longer true. Misleading illustrations are starting to appear in scientific talks. This is due in part to the introduction of business-oriented software in university service courses (usually demanded by the served departments). Errors abound when increased license costs for scientific- and business-oriented software lead departments to eliminate the more scientifically oriented software packages.

The reader should not necessarily interpret these statements as a mandate to avoid business-oriented software. Many of these maligned packages are perfectly capable of producing scientific plots. We must educate ourselves to use the correct software specifications.

Three-dimensional perspective plots are very effective but require specification of a viewpoint. Experiment with various viewpoints to highlight the properties of interest. Mathematical functions lend themselves to three-dimensional surface-type plots, while raw data are typically better illustrated with contour plots. This is especially true for map data, such as surface temperatures or surface wind (where arrows can

denote the direction and the length of the arrow can denote the strength, which effectively adds a fourth dimension of information to the plot).

Figures 10.11 and 10.12 illustrate the population density of children for Harris
County, Texas. Illustration of the data on a map is a natural approach, and a contour
plot reveals the pockets of dense and sparse populations. Further contour plots of
vegetation, topography, roads, and other information may then be sandwiched to
reveal spatial dependencies among various sources of information.

While the contour plot in Figure 10.11 lends itself to comparison of maps, the perspective plot in Figure 10.12 is more difficult to interpret. The surface is more clearly
illustrated, but the surface itself prevents viewing of all of the data.

*Rule 6: Use a contour plot in preference to a perspective plot if a good viewpoint is
not available. Always use a contour plot rather than perspective plot when
the axes denote map coordinates.*

Though the contour plot is generally a better representation of mapped data, the
desire to improve Figure 10.11 would lead us to suggest that the grid lines should
be drawn in a lighter font so that they have less emphasis than the lines for the data
surface. Another improvement to data illustrated according to real-world maps is to
overlay the contour plot where certain known places or geopolitical distinctions
may be marked. The graphic designer must weigh the addition of such decorative
items with the improvement in inference that they bring.

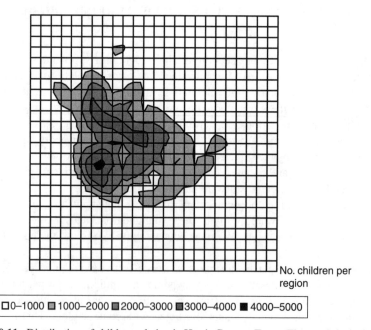

No. children per
region

□0–1000 ■ 1000–2000 ■ 2000–3000 ■ 3000–4000 ■ 4000–5000

Figure 10.11. Distribution of child population in Harris County, Texas. The *x*-axis is the longitude (−96.04 to −94.78 degrees), and the *y*-axis is the latitude (29.46 to 30.26 degrees).

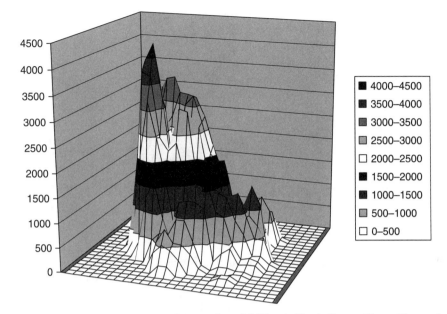

Figure 10.12. Population density of the number of children in Harris County, Texas. The *x*-axis is the longitude (−96.04 to −94.78 degrees), and the *y*-axis is the latitude (29.46 to 30.26 degrees). The *x*-*y* axis is rotated 35 degrees from Figure 10.11.

THE MISUNDERSTOOD AND MALIGNED PIE CHART

The pie chart is undoubtedly the graphical illustration with the worst reputation. Wilkinson [1999] points out that the pie chart is simply a bar chart that has been converted to polar coordinates. Therein lies the problem: most humans naturally think in terms of Cartesian coordinates.

Focusing on Wilkinson's point makes it easier to understand that conversion of the bar height to an angle on the pie chart is most effective when the bar height represents a proportion. If the bars do not have values where the sum of all bars is meaningful, the pie chart is a poor choice for presenting the information (cf. Figure 10.13).

> *Rule 7: Do not use pie charts unless the sum of the entries is scientifically meaningful and of interest to the reader.*

On the other hand, the pie chart is effective for illustrating proportions. This is especially true when we want to focus on a particular slice of the graphic that is approximately 25% or 50% of the data since humans are adept at judging these size portions. Including the actual value as a text element decorating the associated pie slice effectively allows us to communicate the raw number along with the visual clue of the proportion of the total that the category represents. A pie chart intended

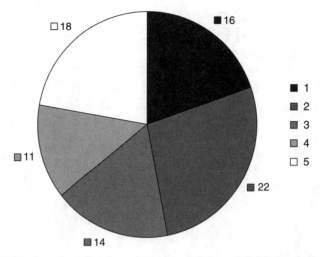

Figure 10.13. Total number of goals scored by Teams 1 through 5. The key indicates the team number and the associated slice color for the number of goals scored by the respective team. The actual number of goals is also included.
Problem: The sum of the individual values is not of interest, so the treatment of the individuals as proportions of a total is not correct.

to display information on all sections where some sections are very small is very difficult to interpret. In these cases, a table or bar chart is preferred.

Additional research has addressed whether the information should be ordered before placement in the pie chart display. There are no general rules other than to repeat that humans are fairly good at identifying pie shapes that are approximately 25% or 50% of the total display. As such, a good ordering of outcomes that included such approximate values would strive to place the leading edge of 25% and 50% pie slices along one of the major north-south or east-west axes. Reordering the set of values may lead to confusion if all other illustrations use a different ordering, so the graphic designer may ultimately feel compelled to reproduce those illustrations as well.

TWO RULES FOR EFFECTIVE DISPLAY OF SUBGROUP INFORMATION

Graphical displays are very effective for communication of subgroup information— for example, when we wish to compare changes in African-American and Hispanic median family income over time. With a moderate number of subgroups, a graphical presentation can be much more effective than a similar tabular display. Labels, stacked bar displays, or a tabular arrangement of graphics can effectively display subgroup information. Each of these approaches has its limits, as we will see in the following sections.

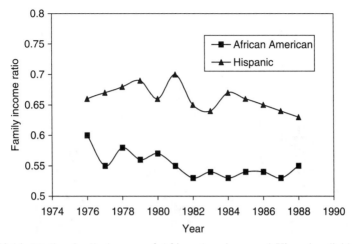

Figure 10.14. Median family income of African-Americans and Hispanics divided by the median family income for Anglo-American families for 1976–1988.
Problem: The legend identifies the two ethnic groups in the reverse order in which they appear in the plot. It is easy to confuse the polylines due to the discrepancy in organizing the identifiers. The rule is that if the data follow a natural ordering in the plotting region, the legend should reflect that order.

In Figure 10.14, separate connected polylines easily separate the subgroup information. Each line is further distinguished with a different plotting symbol. Note how easy it is to confuse the information due to the inverted legend. To avoid this confusion, ensure that the order of entries (top to bottom) matches that of the graphic.

Rule 8: Put the legend items in the same order in which they appear in the graphic whenever possible. You may not know this order until after the graphic has been produced, so check the consistency of this information.

Clearly, there are other illustrations that would work even better for these particular data. When one subgroup is always greater than the other, we can use vertical bars between each measurement instead of two separate polylines. Using data from Table 10.2, a bar chart using subgroups is illustrated in Figure 10.15. Such a display not only points out the discrepancies in the data, but also allows easier inference as to whether the discrepancy is static or changes over time. An improvement in the graphical display appears in Figure 10.16, where more emphasis on the values is achieved by altering the scale of the vertical axis.

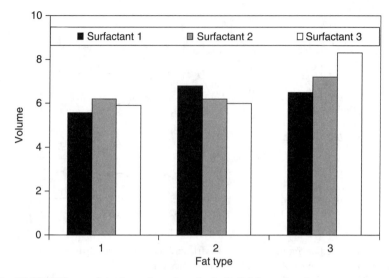

Figure 10.15. Volume of a mixture based on the included fat and surfactant types. *Problem*: As with a scatterplot, the arbitrary decision to include zero on the *y*-axis in a bar plot detracts from the focus on the values plotted.

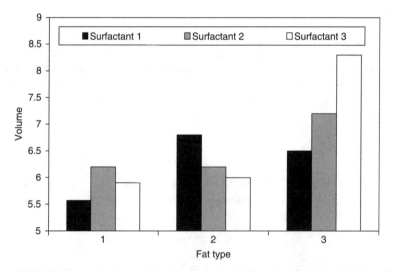

Figure 10.16. Volume of a mixture based on the included fat and surfactant types. Drawing the bar plot with a more reasonable scale clearly distinguishes the values for the reader.

Though the two categorical variables in Table 10.2 are of equal interest, the table uses only one direction for displaying the values of the categories. This reduces the number of dimensions from two to one and makes it more difficult for the reader to discern the subgroup information that the analysis emphasizes.

Tables are simply text-based graphics. All of the rules presented for graphical displays apply equally to textual displays

TABLE 10.2. Volume of a Mixture Based on the Included Fat and Surfactant Types

Fat	Surfactant	Volume
1	1	5.57
1	2	6.20
1	3	5.90
2	1	6.80
2	2	6.20
2	3	6.00
3	1	6.50
3	2	7.20
3	3	8.30

With two categorical variables, the correct approach is to allow one to vary over rows and the other to vary over columns. The organization of Table 10.3 in two dimensions clarifies the subgroup analysis. It is preferable to Table 10.2 and easier to interpret than any graphical representation.

TABLE 10.3. Volume of a Mixture Based on the Included Fat and Surfactant Types

	Surfactant		
Fat	1	2	3
1	5.57	6.20	5.90
2	6.80	6.20	6.00
3	6.50	7.20	8.30

Tables may be augmented with decorative elements just as we augment graphics. Effective additions to the table are judged on their ability to focus attention on the science; otherwise these additions serve as distracters. Specific additions to tables include horizontal and vertical lines to differentiate subgroups, and font/color changes to distinguish headings from data entries.

Specifying a y-axis that starts at zero obscures the differences of the results and violates Rule 3. If we focus on the actual values of the subgroups, we can more readily see the differences.

Rule 9. Use plain language in your legends and text—not "computerese."

An example violating this rule can be seen in the working paper posted at http://www.yuricareport.com/ElectionAftermath04/BerkeleyElection04_WP.pdf, where the authors use the statement "% Democrat Vote Estimated If Electronic Voting = 0" in Place of "Estimated % Vote for Democrats When Printed Ballots Are Used."

TWO RULES FOR TEXT ELEMENTS IN GRAPHICS

If a picture were really worth 1000 words, then graphics would considerably shorten our written reports. While attributing "1000 words" to each graphic is an exaggeration, it remains true that the graphic is often much more efficient at communicating numeric information than equivalent prose. This efficiency concerns the amount of information successfully communicated and not necessarily any space savings.

If the graphic is a summary of numeric information, then the caption is a summary of the graphic. This text element should be considered part of the graphic design and should be carefully constructed rather than added as an afterthought. Readers, for their own use, often copy graphics and tables that appear in articles and reports. The graphic designer's failure to completely document the graphic in the caption can result in gross misrepresentation when the graphic or table is copied and used as a summary in another presentation or report. It is not the presenter who copied the graph who suffers, but the original author who generated the graphic. Tufte [1983] advises that graphics "should be closely integrated with the statistical and verbal descriptions of the data set," and the caption of the graphic clearly provides the best avenue for ensuring this integration.

Rule 10: Captions for your graphical presentations must be complete. Do not skimp on the descriptions.

Do not assume. Though it is common to add a bar representing ± 1.96 standard deviations to some graphs, this addition should be spelled out in the graph's legend or caption, because other graphic designers might use the bar to represent 1 standard deviation. See Tokita et al. [1993] for a particularly flagrant example.

The most effective method for writing a caption is to show the graphic to another person. Allow this person to question the meaning and information presented. Finally, write your explanations as a series of simple sentences for the caption. Readers rarely complain that the caption is too long. If they do, this is a clear indication that the graphic design is poor. If the graphic were more effective, the caption would have a reasonable length.

Depending on the purpose of your report, editors may challenge the duplication of information within the caption and within the text. While we may not win every skirmish with those who want to abbreviate our reports, we are reminded that it is common for others to reproduce only tables and graphics from our reports for other purposes. Detailed captions help reduce misrepresentations and other out-of-context references that we certainly want to avoid. Thus, we endeavor to win as many of these battles with editors as possible.

Other text elements that are important in graphical design are the axis labels, title, and symbols that can be replaced by textual identifiers. Since the plot region of the graph presents numerical data, the axis must declare associated units of measure. If the axis is transformed (log or otherwise), the associated label must present this information as well. The title should be short, serving as a quick reference for the graphic and the associated caption. By itself, the title usually does not contain enough information to fully interpret the graphic in isolation.

When symbols are used to denote points from the data that can be identified by meaningful labels, there are a few choices to consider for improving the information content of the graphic. First, we can replace all symbols with associated labels if this results in a readable (nonoverlapping) presentation. If our focus highlights a few key points, we can substitute labels for only those values.

When replacing (or decorating) symbols with labels results in an overlapping and indecipherable display, a legend is an effective tool, provided that there are not too many legend entries. Producing a graphical legend with 100 entries is not an effective design. It is easy to design these elements when we stop to consider the purpose of the graphic. It is wise to consider two separate graphics when the amount of information overwhelms our ability to document elements in legends and the caption.

Too many line styles or plotting points can be visually confusing and prevent interpretation by the reader. You are better off splitting the single graphic into multiple presentations when there are too many subgroups. A rule of thumb is to limit the number of colors or symbols to seven.

Rule 11: Keep the number of line styles, colors, and symbols to a minimum.

MULTIDIMENSIONAL DISPLAYS

Representing several distinct measures for a collection of points is problematic in both text and graphics. The construction of tables for this display is difficult due to the necessity of effectively communicating the array of subtabular information. The same is true in graphical displays, but the distinction of the various quantities is somewhat easier.

Choosing Effective Display Elements

As Cleveland and McGill [1987] emphasize, graphics involve both encoding of information by the graphic designer and decoding of the information by the reader. Various psychological properties affect information decoding in terms of the reader's graphical perception. For example, when two or more elements are presented, the reader will also envision by-products such as implied texture and shading. These by-products can be distracting and even misleading.

Graphical displays represent the designer's choice concerning the quantitative information that is highlighted. These decisions are based on the desire to assist the analyst and reader in discerning the performance and properties of the data and associated models fitted to the data. While many of the decisions in graphical construction simply follow convention, the designer is still free to choose geometric shapes to represent points, color or style for lines, and shading or textures to represent areas. Cleveland and McGill included a helpful study in which various graphical styles were presented to readers. The ability to discern the underlying information was measured for each style, and an ordered list of effective elementary design choices was inferred. The ordered list for illustrating numeric information is presented in Table 10.4. The purpose of the list is to allow the reader to differentiate among several values.

TABLE 10.4. Rank-Ordered List of Elementary Design Choices for Conveying Numeric Information*

Rank	Graphical element[†]
1	Positions along a common scale
2	Positions along identical nonaligned scales
3	Lengths
4	Angles
4–10	Slopes
6	Areas
7	Volumes
8	Densities
9	Color saturations
10	Color hues

*Slopes are given a wide range of ranks since they can be very poor choices when the aspect ratio of the plot does not allow distinction of slopes. Areas and volumes introduce false dimensions to the display that prevent effective interpretation of the underlying information.
[†]Graphical elements ordered from most (1) to least (10) effective.

When faced with the challenge of depicting a large number of points, there are several steps one should consider when looking for patterns. An interesting challenge was issued by Yi Hui (see http://www.yihui.name/en/category_2.htm). Yi describes generating 20,000 rows (x) and 20,000 columns (y) from a $N(0,1)$ distribution. To this, he also generated 10,000 data points that were on the unit circle ($x^2 + y^2 = 1$). The

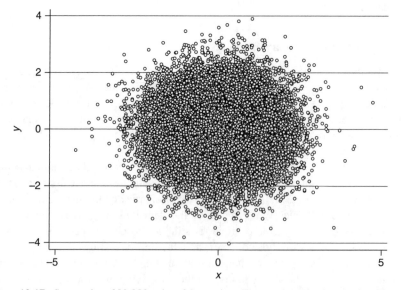

Figure 10.17. Scatterplot of 30,000 pairs of data points. The number of points depicted leads to overlapping hollow circles that don't allow us to see a key feature in the middle of the plot.

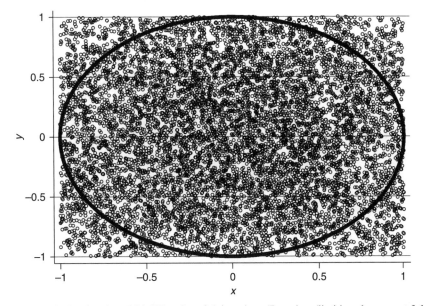

Figure 10.18. Scatterplot of 30,000 pairs of data points. Zooming (limiting the range of the axes) emphasizes the existence of points on the unit circle.

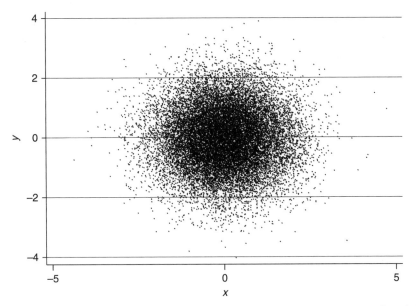

Figure 10.19. Scatterplot of 30,000 pairs of data points. Using a dot rather than a hollow circle for the marker in the plot emphasizes the points on the unit circle (this can be seen better on a computer screen than in this text).

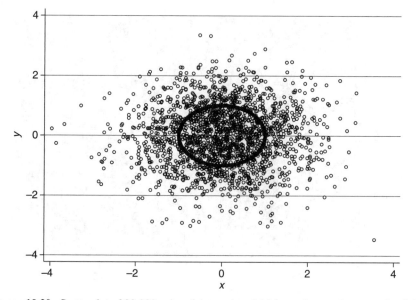

Figure 10.20. Scatterplot of 30,000 pairs of data points. Making only a random sample of the 30,000 pairs of visible points emphasizes the points on the unit circle.

description of the data should be enough to allow the interested reader to generate a data set with two variables (x and y) with 30,000 observations.

The challenge is to draw a scatterplot which reveals the circle pattern (the 10,000 points that are on the unit circle). An initial plot for which small circles denote each pair is simply too dark (overlapping circles) in the middle of the illustration to allow one to see that there are a number of observations on the unit circle; see Figure 10.17.

There are various approaches to consider when trying to find a pattern in a large amount of data. In the first approach, we zoom in on the large amount of information by limiting the axes. This approach is seen in Figure 10.18. A second approach is to draw all of the data but reduce the symbol to a single dot. This approach works better on a computer screen (especially one which allows us to make the overall picture larger) than it does on a piece of paper; see Figure 10.19. Finally, not knowing where among the data points a feature may be hidden, we draw a small random sample of the data to see if any pattern appears in Figure 10.20.

CHOOSING GRAPHICAL DISPLAYS

When relying completely on the ability of software to produce scientific displays, many authors are limited by their mastery of the software. Most software packages will allow users either to specify in advance the desired properties of the graph or to edit the graph to change individual items. Our ability to follow the guidelines outlined in this chapter is directly related to the time we spend learning to use the more advanced graphics features of software.

Oral Presentations

Graphs. The rules for graphics in print are equally applicable to lectures and may be summed up as follows: "Never use a chart that will take longer to explain than the information it was intended to provide."

Tables. The numeric values in a table should occupy no more than three columns and include no more than three digits each, for example, 318, 3.18, 3.1×10^8.

Text. A slide should contain no more than three bullet points, as in Figure 10.21a, and should *never* be merely a rehash of the lecture itself, as in Figure 10.21b.

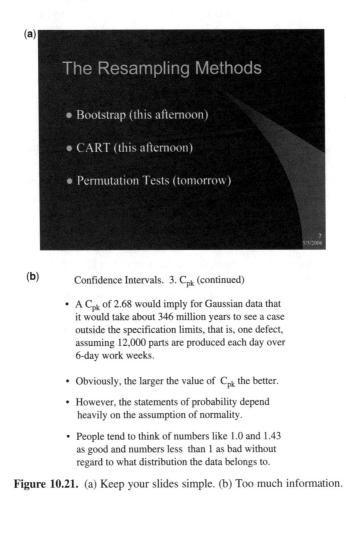

(a)

The Resampling Methods

- Bootstrap (this afternoon)

- CART (this afternoon)

- Permutation Tests (tomorrow)

7
5/5/2008

(b)   Confidence Intervals. 3. C_{pk} (continued)

- A C_{pk} of 2.68 would imply for Gaussian data that it would take about 346 million years to see a case outside the specification limits, that is, one defect, assuming 12,000 parts are produced each day over 6-day work weeks.

- Obviously, the larger the value of C_{pk} the better.

- However, the statements of probability depend heavily on the assumption of normality.

- People tend to think of numbers like 1.0 and 1.43 as good and numbers less than 1 as bad without regard to what distribution the data belongs to.

Figure 10.21. (a) Keep your slides simple. (b) Too much information.

SUMMARY

- Examine the data and the results to determine the number of dimensions in the information to be illustrated. Limit your graphic to that number of dimensions.

- Limit the axes to exactly (or closely) match the range of data in the presentation unless a zero axis limit admits desired relative comparisons of the depicted values.

- Do not connect points in a scatterplot unless there is an underlying interpolation that makes scientific sense.

- Recognize that readers of your reports will copy tables and figures for their own use. Ensure that you are not misquoted by completely describing your graphics and tables in the associated legends. Do not skimp on these descriptions or you will force readers to scan the entire document for needed explanations.

- If readers are to accurately compare two different graphics for values (instead of shapes or predominant placement of outcomes), use the same axis ranges on the two plots.

- Use pie charts only when there is a small number of categories and the sum of the categorical values has scientific meaning.

- Tables are text-based graphics. Therefore, the rules governing the organization and scientific presentation of graphics should be followed for the tables that we present. Headings should be differentiated from data entries by font weight or color change. Refrain from introducing multiple fonts in the tables; instead, use one font where differences are denoted by weight (boldness), style (slanted), and size.

- Numeric entries in tables should have the same number of significant digits. Further, they should be right justified so that they line up and allow easy interpretation while scanning columns of numbers.

- Many charts can benefit from the addition of grid lines. Bar charts especially can benefit from horizontal grid lines from the y-axis labels. This is especially true of wider displays, but grid lines should be drawn in a lighter shade than the lines used to draw the major features of the graphic.

- Critique your graphics and tables after production by isolating them with their associated captions. Determine if the salient information is obvious by asking a colleague to interpret the display. If we are serious about producing efficient communicative graphics, we must take the time to ensure that our graphics are interpretable.

TO LEARN MORE

For many more examples of bad and/or misleading graphics, see http://www.math.yorku.ca/SCS/Gallery/. Wilkinson [1999] presents a formal grammar for describing graphics but, more importantly (for our purposes), the author lists graphical element hierarchies from best to worst. Cleveland [1994] focuses on the elements of common illustrations, exploring the effectiveness of each element in communicating numeric information. A classic text is Tukey [1977], which lists both graphical and text-based graphical summaries of data. Tufte [1983, 1990] organized much of the previous work and combined that work with modern developments. See also Burn [1993] and Wainer [1997, 2004]. For specific illustrations, subject-specific text books can be consulted for particular displays in context; Hardin and Hilbe [2003, pp. 143–167] illustrate the use of graphics for assessing model accuracy.

PART IV

BUILDING A MODEL

11

UNIVARIATE REGRESSION

Are the data adequate? Does your data set cover the entire range of interest? Will your model depend on one or two isolated data points?

The simplest example of a model, the relationship between exactly two variables, illustrates at least five of the many complications that can interfere with the task of model building:

1. Limited scope—the model we develop may be applicable for only a portion of the range of each variable.
2. Ambiguous form of the relationship—a variable may give rise to a statistically significant linear regression without the underlying relationship being a straight line.
3. Confounding—undefined confounding variables may create the illusion of a relationship or may mask an existing one.
4. Assumptions—the assumptions underlying the statistical procedures we use may not be satisfied.
5. Inadequacy—goodness of fit is not the same as prediction.

We consider each of these error sources in turn along with a series of preventive measures. Our discussion is divided into problems connected with model selection and difficulties that arise during the estimation of model coefficients.

Common Errors in Statistics (and How to Avoid Them), Third Edition. Edited by P. I. Good and J. W. Hardin
Copyright © 2009 John Wiley & Sons, Inc.

MODEL SELECTION

Limited Scope

Almost every relationship has both a linear and a nonlinear portion with the nonlinearities becoming more evident as we approach the extremes of the independent (causal) variable's range. One can think of many examples from physics such as Boyles Law, which fails at high pressures, and particle symmetries that are broken as the temperature falls.

Almost every measuing device—electrical, electronic, mechanical, or biological, is reliable only in the central portion of its scale. In medicine, radioimmune assay fails to deliver reliable readings at very low dilutions; this has practical implications, as an increasing proportion of patients will fail to respond as the dosage drops.

We need to recognize that while a regression equation may be used for interpolation within the range of measured values, we are on shaky ground if we try to extrapolate, to make predictions for conditions not previously investigated. The solution is to know the range of application and to recognize, even if we do not exactly know the range, that our equations will be applicable to some but not all possibilities.

Ambiguous Relationships

Think why rather than what.

The exact nature of the formula connecting two variables cannot be determined by statistical methods alone. If a linear relationship exists between two variables X and Y, then a linear relationship also exists between Y and any monotone (nondecreasing or nonincreasing) function of X. Assume that X can only take positive values. If we can fit Model I: $Y = \alpha + \beta X + \varepsilon$, to the data, we also can fit Model II: $Y = \alpha' + \beta' \log[X] + \varepsilon$, and Model III: $Y = \alpha'' + \beta'' X + \gamma X^2 + \varepsilon$. It can be very difficult to determine which model, if any, is the correct one in either a predictive or a mechanistic sense.

A graph of Model I is a straight line (see Fig. 11.1). Because Y includes a stochastic or random component ε, the pairs of observations (x_1, y_1), (x_2, y_2), ... will not fall exactly on this line but above and below it. The function $\log[X]$ does not increase as rapidly as X does; when we fit Model II to these same pairs of observations, its graph rises above that of Model I for small values of X and falls below that of Model I for large values. Depending on the set of observations, Model II may give just as good a fit to the data as Model I.

How Model III behaves will depend upon whether β'' and α'' are both positive or whether one is positive and the other negative. If β'' and α'' are both positive, then the graph of Model III will lie below the graph of Model I for small positive values of X and above it for large values. If β'' is positive and α'' is negative, then Model III will behave more like Model II. Thus, Model III is more flexible than either Model I or II and can usually be made to give a better fit to the data, that is, to minimize some function of the differences between what is observed, y_i, and what is predicted by the model, $Y[x_i]$.

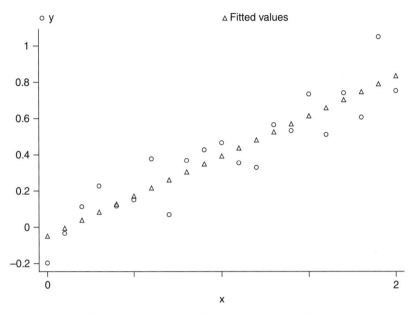

Figure 11.1. A straight line appears to fit the data.

The coefficients α, β, γ for all three models can be estimated by a technique known (to statisticians) as linear regression. Our knowledge of this technique should not blind us to the possibility that the true underlying model may require nonlinear estimation, as in

$$\text{Model IV: } Y = \frac{\alpha + \beta X + \gamma X^2}{\delta - \phi X} + \varepsilon$$

This latter model may have the advantage over the first three in that it fits the data over a wider range of values.

Which model should we choose? At least two contradictory rules apply:

- The more parameters the better the fit; thus, Models III and IV are to be preferred.
- The simpler, more straightforward model is more likely to be correct when we come to apply it to data other than the observations in hand; thus, Models I and II are to be preferred.

Again, the best rule of all is not to let statistics do your thinking for you, but to inquire into the mechanisms that give rise to the data and that might account for the relationship between variables X and Y. An example taken from physics is the relationship between the volume V and temperature T of a gas. All of the preceding four models could be used to fit the relationship. But only one, the model $V = \alpha + KT$, is consistent with kinetic molecular theory.

Inappropriate Models

An example in which the simpler, more straightforward model is not correct occurs when we try to fit a straight line to what is actually a higher-order polynomial. For example, suppose we tried to fit a straight line to the relationship $Y = (X - 1)^2$ over the range $X = (0, +2)$. We'd get a line with slope 0, similar to that depicted in Figure 11.2. With a correlation of 0, we might even conclude in error that X and Y are not related. Figure 11.2 suggests a way we can avoid falling into a similar trap.

Always plot the data before deciding on a model.

The data in Figure 11.3 are taken from Mena et al. [1995]. These authors reported in their abstract that "The correlation ... between IL-6 and TNF-alpha was .77 ... statistically significant at a p-value less than .01." Would you have reached the same conclusion?

With more complicated models, particularly those like Model IV that are nonlinear, it is advisable to calculate several values that fall outside the observed range. If the results appear to defy common sense (or the laws of physics, market forces, etc.), the nonlinear approach should be abandoned and a simpler model utilized.

Often it can be difficult to determine which variable is the cause and which is the effect. But if the values of one of the variables are fixed in advance, then this variable should always be treated as the so-called independent variable or cause, the X in the equation $Y = a + bX + \varepsilon$. Here is the reason.

When we write $Y = a + bx + \varepsilon$, we actually mean, $Y = E(Y \mid x) + \varepsilon$, where $E(Y \mid X) = a + bx$ is the expected value of an indefinite number of independent

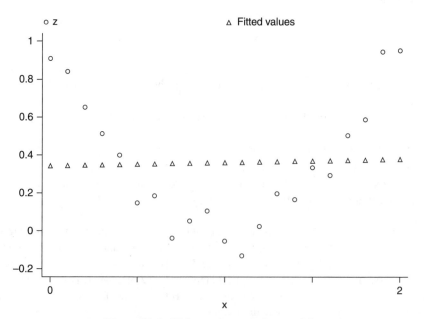

Figure 11.2. Fitting an inappropriate model.

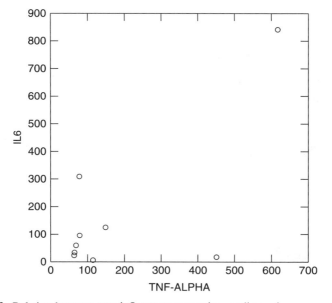

Figure 11.3. Relation between two inflammatory reaction mediators in response to silicone exposure. Data taken from Mena et al. [1995].

observations of Y when $X = x$. If X is fixed, the inverse equation $x = (E(x \mid Y) - a)/b + \varepsilon' =$ makes little sense.

Nonuniqueness

Though a model may provide a good fit to a set of data, one should refrain from inferring any causal connection. The reason is that a single model is capable of fitting many disparate data sets. Consider that one line, $Y = 3 + 0.5X$, fits the four sets of paired observations depicted in Figures 11.4a–d with $R^2 = 0.67$ in each case.

The data for these four figures are as follows:

$X1 = c(10,8,13,9,11,14,6,4,12,7,5)$
$X2 = c(8,8,8,8,8,8,8,19,8,8,8)$
$Y1 = c(8.04,6.95,7.58,8.81,8.33,9.96,7.24,4.26,10.84,4.82,5.68)$
$Y2 = c(9.14,8.14,8.74,8.77,9.26,8.10,6.13,3.10,9.13,7.26,4.74)$
$Y3 = c(7.46,6.77,12.74,7.11,7.81,8.84,6.08,5.39,8.15,6.42,5.73)$
$Y4 = c(6.58,5.76,7.71,8.84,8.47,7.04,5.25,12.50,5.56,7.91,6.89)$

Confounding Variables

If the effects of additional variables other than X on Y are suspected, these additional effects should be accounted for either by stratifying or by performing a multivariate regression.

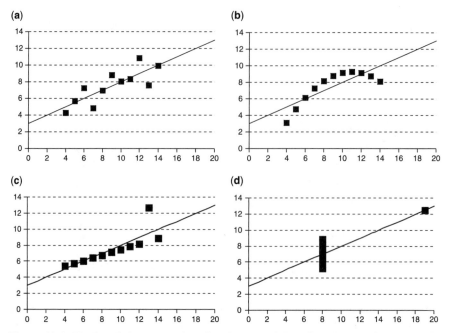

Figure 11.4. The best-fitting (regression) line for each of these four data sets is $y = 3 + .5x$, where each regression is characterized by $R^2 = 0.67$. Plot (a) is a reasonable data set for which the linear model is applied. Plot (b) illustrates a possible quadratic relationship not accounted for by our linear model. Plot (c) demonstrates the effect of a possibly miscoded outcome value yielding a slope that is somewhat higher than it otherwise would be, while plot (d) demonstrates an outlier with very high leverage.

SOLVE THE RIGHT PROBLEM

Don't be too quick to turn on the computer. Bypassing the brain to compute by reflex is a sure recipe for disaster.

Be sure of your objectives for the model: Are you trying to uncover cause-and-effect mechanisms? Or derive a formula for use in predictions? If the former is your objective, standard regression methods may not be appropriate.

A researcher studying how neighborhood poverty levels affect violent crime rates hit an apparent statistical roadblock. Some important criminological theories suggest that this positive relationship is curvilinear with an accelerating slope, while other theories suggest a decelerating slope. As the crime data are highly variable, previous analysts had used the logarithm of the primary end point—a violent crime rate—and reported a significant negative quadratic term (poverty * poverty) in their least squares models. The researcher felt that such results were suspect, that the log transformation alone might have biased the results toward finding a significant negative quadratic term for poverty.

But quadratic terms and log transforms are irrelevancies, artifacts resulting from an attempt to squeeze the data into the confines of a linear regression model. The issue appears to be whether the rate of change of crime rates with poverty levels is a constant, increasing, or decreasing function of poverty levels. Resolution of this issue requires a totally different approach.

Suppose Y is the variable you are trying to predict and X is the predictor. Replace each of the $y[i]$ by the slope $y^*[i] = (y[i+1] - y[i])/(x[i+1] - x[i])$. Replace each of the $x[i]$ by the midpoint of the interval over which the slope is measured, $x^*[i] = (x[i+1] - x[i])/2$. Use the permutation methods described in Chapter 5 to test for the correlation, if any, between y^* and x^*. A positive correlation means an accelerating slope, a negative correlation a decelerating slope.

Correlations can be deceptive. Variable X can have a statistically significant correlation with variable Y solely because X and Y are both dependent on a third variable, Z. A fall in the price of corn is inversely proportional to the number of hay fever cases only because the weather that produces a bumper crop of corn generally yields a bumper crop of ragweed as well.

Even if the causal force X under consideration has no influence on the dependent variable Y, the effects of unmeasured selective processes can produce an apparent test effect. Children were once taught that storks brought babies. This juxtaposition of bird and baby makes sense (at least to a child), for where there are houses, there are both families and chimneys where storks can nest. The bad air or miasma model (common sense two centuries ago) works rather well at explaining respiratory illnesses and not at all at explaining intestinal ones. An understanding of the roles that bacteria and viruses play unites the two types of illness and enriches our understanding of both.

We often try to turn such pseudocorrelations to advantage in our research, using readily measured *proxy variables* in place of their less easily measured causes. Examples are our use of population change in place of economic growth, M2 for the desire to invest, arm cuff blood pressure measurement in place of the width of the arterial lumen, and tumor size for mortality. At best, such *surrogate responses* are inadequate (as in attempting to predict changes in stock prices); at worst, they may actually point in the wrong direction.

At one time, the level of CD-4 lymphocytes in the blood appeared to be associated with the severity of AIDs; the result was that a number of clinical trials used changes in this level as an indicator of disease status. Reviewing the results of 16 sets of such trials, Fleming [1995] found that CD-4 rose to favorable levels in 13 instances even though the clinical outcomes were favorable in only 8.

STRATIFICATION

Gender discrimination lawsuits based on the discrepancy in pay between men and women could be defeated once it was realized that pay was related to years in service and that women who had only recently entered the job market in great numbers simply didn't have as many years on the job as men.

These same discrimination lawsuits could be won once the gender comparison was made on a years-in-service basis, that is, when the salaries of new female employees were compared with those of newly employed men, the salaries of women with three years of service with those of men with the same time in grade, and so forth. Within each stratum, men always had the higher salaries.

If the effects of additional variables other than X on Y are suspected, they should be accounted for either by stratifying or by performing a multivariate regression as described in the next chapter.

The two approaches are *not* equivalent unless *all* terms are included in the multivariate model. Suppose we want to account for the possible effects of gender. Let $I[]$ be an indicator function that takes the value 1 if its argument is true and 0 otherwise. Then, to duplicate the effects of stratification, we would have to write the multivariate model in the following form:

$$Y = a_m I[\text{male}] + a_f(1 - I[\text{male}]) + b_m I[\text{male}]X + b_f(1 - I[\text{male}]) + e$$

In a study by Kanarek et al. [1980], whose primary focus is the relation between asbestos in drinking water and cancer, results are stratified by sex, race, and census tract. Regression is used to adjust for income, education, marital status, and occupational exposure.

Lieberson [1985] warns that if the strata differ in the levels of some third unmeasured factor that influences the outcome variable, the results may be bogus.

Simpson's Paradox

A third omitted variable may also result in two variables appearing to be independent when the opposite is true. Consider the following table, an example of what is termed Simpson's paradox:

	Treatment Group	
	Control	Treated
Alive	6	20
Dead	6	20

We don't need a computer program to tell us that the treatment has no effect on the death rate. Or does it? Consider the following two tables that result when we examine males and females separately:

	Males	
	Control	Treated
Alive	4	8
Dead	3	5

	Females	
	Control	Treated
Alive	2	12
Dead	3	15

In the first of these tables, treatment reduces the male death rate from 3 out of 7, or 0.43, to 5 out of 13, or 0.38. In the second table, treatment reduces the female death rate from 3 out of 5, or 0.6, to 15 out of 27, or 0.55. Both sexes show a reduction, yet the combined population does not. This paradox is resolved by avoiding a knee-jerk response to statistical significance when association is involved. One needs to think deeply about the underlying cause-and-effect relationships before analyzing the data. Thinking about cause and effect in the preceding example might have led us to think about possible sexual differences and to stratifying the data by sex before analyzing them.

ESTIMATING COEFFICIENTS

Write down and confirm your assumptions before you begin.

In this section we consider problems and solutions associated with three related challenges:

1. Estimating the coefficients of a model.
2. Testing hypotheses concerning the coefficients.
3. Estimating the precision of our estimates.

The techniques we employ will depend upon the following:

1. The nature of the regression function (linear, nonlinear, logistic).
2. The nature of the losses associated with applying the model.
3. The distribution of the error terms in the model, that is, the εs.
4. Whether these error terms are independent or dependent.

The estimates we obtain will depend upon our choice of fitting function. This choice should not be dictated by the software but by the nature of the losses associated with applying the model. Our software may specify a least-squares fit—most commercially available statistical packages do—but our real concern may be with minimizing the sum of the absolute values of the prediction errors or the maximum loss to which one will be exposed. A solution is provided in the next chapter.

In the *univariate* linear regression model, we assume that

$$y = E(Y \mid x) + \varepsilon$$

where E denotes the mathematical expectation of Y given x and could be any deterministic function of x in which the parameters appear in linear form. ε, the error term, stands for all the other unaccounted-for factors that make up the observed value y.

How accurate our estimates are and how consistent they will be from sample to sample will depend upon the nature of the error terms. If none of the many factors that contribute to the value of ε make more than a small contribution to the total, then ε will have a Gaussian distribution. If the $\{\varepsilon_i\}$ are independent and normally distributed (Gaussian), then the ordinary least squares estimates of the coefficients produced by most statistical software will be unbiased and have minimum variance.

These desirable properties, indeed the ability to obtain coefficient values that are of use in practical applications, will not be present if the wrong model has been adopted. They will not be present if successive observations are dependent. The values of the coefficients produced by the software will not be of use if the associated losses depend on some function of the observations other than the sum of the squares of the differences between what is observed and what is predicted. In many practical problems, one is more concerned with minimizing the sum of the absolute values of the differences or with minimizing the maximum prediction error. Finally, if the error terms come from a distribution that is far from Gaussian, a distribution that is truncated, flattened, or asymmetric, the p-values and precision estimates produced by the software may be far from correct.

Alternatively, we may use permutation methods to test for the significance of the resulting coefficients. Provided that the $\{\varepsilon_i\}$ are i.i.d. (Gaussian or not), the resulting p-values will be exact. They will be exact regardless of which goodnessof-fit criterion is employed.

Suppose that our hypothesis is that $y_i = a + bx_i + \varepsilon_i$ for all i and $b = b_o$. First, we substitute $y_i' = y_i - b_o x_i$ in place of the original observations y_i. Our translated hypothesis is $y_i' = a + b'x_i + \varepsilon_i$ for all i and $b' = 0$ or, equivalently, $\rho = 0$, where ρ is the correlation between the variables Y' and X. Our test for correlation is based on the permutation distribution of the sum of the cross-products $y_i'x_i$ [Pitman, 1938]. Alternative tests based on permutations include those of Cade and Richards [1996] and Multi Response Permutation Procedure LAD regression [Mielke and Berry, 1997].

For large samples, these tests are every bit as sensitive as the least-squares test described in the previous paragraph even when all the conditions for applying that test are satisfied [Mielke and Berry, 2001, Section 5.4].

If the errors are dependent and normally distributed and the covariances are the same for every pair of errors, then we may also apply any of the permutation methods described above. If the errors are dependent and normally distributed, but we are reluctant to make such a strong assumption about the covariances, then our analysis may call for dynamic regression models [Pankratz, 1991].[1]

---

[1]In the SAS manual, these are called ARIMAX techniques and are incorporated in Proc ARIMA.

FURTHER CONSIDERATIONS

Bad Data

The presence of bad data can completely distort regression calculations. When least-squares methods are employed, a single outlier can influence the entire line to pass closely to the outlier. While a number of methods exist for detecting the most influential observations [e.g., Mosteller and Tukey, 1977], influence does not automatically mean that the data point is in error. Measures of influence encourage review of data for exclusion. Statistics do not exclude data; analysts do. And they exclude data only when presented with firm evidence that the data are in error.

The problem of bad data is particularly acute in two instances:

1. When most of the data are at one end of the line, so that a few observations at the far end can have undue influence on the estimated model.
2. When there is no causal relationship between X and Y.

The Washington State Department of Social and Health Services extrapolates its audit results on the basis of a regression of over- and undercharges against the dollar amount of the claim. As the frequency of errors depends on the amount of paperwork involved and not on the dollar amount of the claim, no linear relationship exists between overcharges and the amount of the claim. The slope of the regression line can vary widely from sample to sample; removal from or addition of a very few samples to the original audit can dramatically affect the amount claimed by the state in overcharges.

Recommended is the *delete-one* approach, in which the regression coefficients are recomputed repeatedly, deleting a single pair of observations from the original data set each time. These calculations provide confidence intervals for the estimates along with an estimate of the sensitivity of the regression to outliers. When the number of data pairs exceeds 100, a bootstrap might be used instead.

> To get an estimate of the precision of the estimates and the sensitivity of the regression equation to bad data, recompute the coefficients, leaving out a different data pair each time.

Convenience

More often than we would like to admit, the variables and data that go into our models are chosen for us. We cannot directly measure the variables we are interested in, so we make do with surrogates. But such surrogates may or may not be directly related to the variables of interest. Lack of funds and or/the necessary instrumentation limits the range over which observations can be made. Our census overlooks the homeless, the uncooperative, and the less luminous. [See, for example, *City of New York v. Dept. of Commerce*;[2] Bothun, 1998, Chapter 6.]

[2]822 F. Supp. 906 (E.D.N.Y., 1993).

The presence of such bias does not mean that we should abandon our attempts at modeling, but rather that we should be aware of and report our limitations.

WILL WOMEN RUNNERS EVER OVERTAKE MEN AT THE OLYMPICS?

In an article deliberately designed to provoke controversy, A. J. Tatem and colleagues [2004] suggested that women sprinters may one day overtake men. They began their demonstration by fitting linear regression lines to the best times recorded in the Olympics from 1900 to 2004. Then they extrapolated these lines well into the twenty-second century. Critics raised numerous objections; see *Nature* [2004, 132, p. 137], the most obvious being that if their results are extended in a purely linear fashion to the twenty-seventh century, times of less than zero seconds were sure to be recorded.

Using the best 10 times each year rather than the best time at each Olympiad yields 40 times as much data and reveals several breakpoints in the "linear" curves. One resulted from an increase in the number of women competing, another from increases in the number of training sessions. The latter has already reached a plateau. See www.antenna.nl/weia/Progressie.html.

Stationarity

An underlying assumption of regression methods is that relationships among variables remain constant during the data collection period. If not, if the variables we are measuring undergo seasonal or other detectable changes, then we need to account for them. A multivariate approach is called for, as described in the next chapter.

Practical versus Statistical Significance

An association can be of statistical significance without being of the least practical value. In the study by Kanarek et al. [1980] referenced above, a 100-fold increase in asbestos fiber concentration is associated with perhaps a 5% increase in lung cancer rates. Do we care? Perhaps, for no life can be considered unimportant. But courts traditionally have looked for at least a twofold increase in incidence before awarding damages. [See, for example, the citations in Chapter 6 of Good, 2001.] And in this particular study, there is reason to believe that there might be other hidden cofactors that are at least as important as the presence of asbestos fiber.

Goodness of Fit versus Prediction

As noted above, we have a choice of fitting methods: We can minimize the sum of the squares of the deviations between the observed and model values, or we can minimize the sum of the absolute values of these deviations, or we can minimize some entirely different function. Suppose that we have followed the advice given above and have chosen our goodness-of-fit criterion to be identical with our loss function.

For example, suppose the losses are proportional to the square of the prediction errors, and we have chosen our model's parameters so as to minimize the sum of squares of the differences $y_i - M[x_i]$ for the historical data. Unfortunately, minimizing this sum of squares is no guarantee that when we continue to make observations, we will continue to minimize the sum of squares between what we observe and what our model predicts. If you are a businessperson whose objective is to predict market response, this distinction can be critical.

There are at least three reasons for the possible disparity:

1. The original correlation was spurious.
2. The original correlation was genuine but the sample was not representative.
3. The original correlation was genuine but the nature of the relationship has changed with time. (As a result of changes in the underlying political culture, economy, or environment, for example.) We take up this problem again in Chapter 15.

And lest we forget: association does not prove causation, it only contributes to the evidence.

Indicator Variables

The use of an indicator (yes/no) or a nonmetric ordinal variable (improved, much improved, no change) as the sole independent (X) variable is inappropriate. The two-sample and k-sample procedures described in Chapter 5 should be employed.

Transformations

It is often the case that the magnitude of the residual error is proportional to the size of the observations, that is, $y = E(Y \mid x)\varepsilon$. A preliminary log transformation will restore the problem to linear form, $\log(y) = \log E(Y \mid x) + \varepsilon'$. Unfortunately, even if ε is normal, ε' is not, and the resulting confidence intervals need be adjusted (Zhou and Gao, 1997).

When a Straight Line Won't Do

Few processes are purely linear. Almost all have an S shape, though the lengths of the left and right horizontals of the S may differ. Sometimes, the S shape results from the measuring instruments, which typically fail for very large or very small values. But equally often, it is because there is a lower threshold that needs be overcome and an upper limit that results from saturation.

For example, in Figure 11.5a, the linear relationship between the dependent variable Y and the time t appears obvious. But as seen in Figure 11.5b, the actual relationship between Y and t is that of logistic growth.

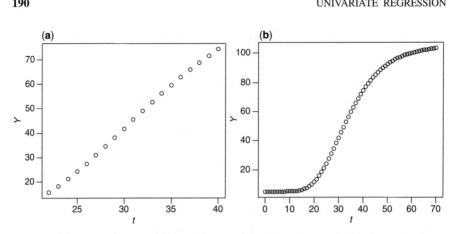

Figure 11.5. (a) The linear relationship between the dependent variable Y and the time t appears obvious. (b) But the actual relationship between Y and t is that of logistic growth.

We have already distinguished processes where growth is additive with time, $y = a + bt$, from those where it is multiplicative, $y = ae^{bt}$, and we work instead with the relationship $\log[y] = a + bt$. But the growth of welfare in the 1960s was found to occur in four phases:

1. First, the growth was additive as recipients drifted into the program at random. Written as a differential equation, this would be $dy/dt = b$.
2. Then a multiplicative component was added as the knowledge of welfare spread from current recipients to potential ones: $dy/dt = b + cy$ or $\log[y] = b + ct$.
3. When recipients began to organize and actively recruit other recipients, the relationship took the form $dy/dt = b + cy + fy^2$.
4. Finally, almost everyone who was eligible for welfare was receiving it, and the growth of the program more closely resembled a logistic curve with $dy/dt = (1 - y/K)(b + cy + fy^2)$.

Curve Fitting and Magic Beans

Until recently, what distinguished statistics from the other branches of mathematics was that at least one aspect of each analysis was firmly grounded in reality. Samples were drawn from real populations and in theory, one could assess and validate findings by examining larger and larger samples taken from that same population.

In this reality-based context, modeling has one or possibly both of the following objectives:

1. To better understand the mechanisms leading to particular responses.
2. To predict future outcomes.

Failure to achieve these objectives has measurable losses. While these losses cannot be eliminated because of the variation inherent in the underlying processes, hopefully, by use of the appropriate statistical procedure, they can be minimized.

By contrast, the goals of curve fitting (nonparametric or local regression)[3] are aesthetic in nature; the resultant graphs, though pleasing to the eye, may bear little relation to the processes under investigation. To quote Green and Silverman [1994, p. 50], "there are two aims in curve estimation, which to some extent conflict with one another, to maximize goodness-of-fit and to minimize roughness."

The first of these aims is appropriate *if goodness of fit means minimizing* the loss function.[4] The second creates a strong risk of overfitting.

Validation is essential, yet most of the methods discussed in Chapter 14 do not apply. Validation via a completely independent data set cannot provide confirmation, as the new data would entail the production of a completely different, unrelated curve. The only effective method of validation is to divide the data set in half at random, fit a curve to one of the halves, and then assess its fit against the entire data set.

SUMMARY

Regression methods work well with physical models. The relevant variables are known, and so are the functional forms of the equations connecting them. Measurement can be done to high precision, and much is known about the nature of the errors—in the measurements and in the equations. Furthermore, there is ample opportunity for comparing predictions to reality.

Regression methods can be less successful for biological and social science applications. Before undertaking a univariate regression, you should have a fairly clear idea of the mechanistic nature of the relationship (and thus the form the regression function will take). Look for deviations from the model, particularly at the extremes of the variable range. A plot of the residuals can be helpful in this regard; see, for example, Davison and Snell [1991] and Hardin and Hilbe [2003, pp. 143–159].

A preliminary multivariate analysis (the topic of the next two chapters) will give you a fairly clear notion of which variables are likely to be confounded so that you can correct for them by stratification. Stratification will also allow you to take advantage of permutation methods which are to be preferred in instances where errors or model residuals are unlikely to follow a normal distribution.

It's also essential that you have firmly in mind the objectives of your analysis, and the losses associated with potential decisions, so that you can adopt the appropriate method of goodness of fit. The results of a regression analysis should be treated with care; as Freedman [1999] notes, "Even if significance can be determined and the null hypothesis rejected or accepted, there is a much deeper problem. To make causal inferences, it must in essence be assumed that equations are invariant under proposed interventions.... [I]f the coefficients and error terms change when the variables on the right-hand side of the equation are manipulated rather than being

---

[3]See, for example, Green and Silverman [1994] and Loader [1999].
[4]Most published methods also require that the loss function be least-squares and the residuals be normally distributed.

passively observed, then the equation has only a limited utility for predicting the results of interventions."

Statistically significant findings should serve as a motivation for further corroborative and collateral research rather than as a basis for conclusions.

Checklist: Write Down and Confirm Your Assumptions Before You Begin

- Data cover an adequate range. The slope of the line is not dependent on a few isolated values.
- The model is plausible and has or suggests a causal basis.
- Relationships among variables remained unchanged during the data collection period and will remain unchanged in the near future.
- Uncontrolled variables are accounted for.
- The loss function is known and will be used to determine the goodness-of-fit criteria.
- Observations are independent, or the form of the dependence is known or is a focus of the investigation.
- The regression method is appropriate for the types of data involved and the nature of the relationship.
- The distribution of residual errors is known.

TO LEARN MORE

David Freedman's [1999] article on association and causation is essential reading. Lieberson [1985] presents many examples of spurious association. Friedman, Furberg and DeMets [1996] cite a number of examples of clinical trials using misleading surrogate variables.

Mosteller and Tukey [1977] expand on many of the points raised here concerning the limitations of linear regression. Distribution-free methods for comparing regression lines among strata are described by Good [2001, pp. 168–169].

For more on Simpson's paradox, see http://www.cawtech.freeserve.co.uk/simpsons.2.html. For a real-world example, search under Simpson's paradox for an analysis of racial bias in New Zealand jury service at http://www.stats.govt.nz.

12

ALTERNATE METHODS OF REGRESSION

"Imagine how statisticians might feel about the powerful statistics programs that are now in our hands. It is so easy to key-in a set of data and calculate a wide variety of statistics—regardless what those statistics are or what they mean. There also is a need to check that things are done correctly in the statistical analyses we perform in our laboratories."

—James O. Westgard [1998]

In the previous chapter, we focused exclusively on *ordinary least-squares* linear regression (OLS) both because it is the most common modeling technique and because the limitations and caveats we outlined there apply to virtually all modeling techniques. But OLS is not the only modeling technique.

If one wants to diminish the effect of outliers, and treat prediction errors as proportional to their absolute magnitude rather than their squares, one should use *least absolute deviation* (LAD) regression.

If it is not clear which variable should be viewed as the predictor and which the dependent variable, as is the case when evaluating two methods of measurement, then one should employ Deming or *error in variable* (EIV) regression.

If one's primary interest is not in the expected value of the dependent variable, but in its extremes (the number of bacteria that will survive treatment, the number of individuals who will fall below the poverty line), then one ought consider the use of *quantile regression.*

If distinct strata exist, one should consider developing separate regression models for each stratum, a technique known as ecological regression, discussed in the next-to-last section of the present chapter.

If one's interest is in classification or if the majority of one's predictors are dichotomous, then one should consider the use of *Classification and Regression Tress* (CART) discussed in the next chapter.

Common Errors in Statistics (and How to Avoid Them), Third Edition. Edited by P. I. Good and J. W. Hardin
Copyright © 2009 John Wiley & Sons, Inc.

If the outcomes are limited to success or failure, one ought employ *logistic regression*. If the outcomes are counts rather than continuous measurements, one should employ a *generalized linear model* (GLM). See Chapter 13.

LINEAR VERSUS NON-LINEAR REGRESSION

Linear regression is a much misunderstood and mistaught concept. If a linear model provides a good fit to data, this does not imply that a plot of the dependent variable with respect to the predictor would be a straight line, only that a plot of the dependent variable with respect to some not necessarily monotonic function of the predictor would be a straight line.

For example, $y = A + B \log[x]$ and $y = A \cos(x) + B \sin(x)$ are both linear models whose coefficients A and B might be derived by OLS or LAD methods. $Y = Ax^5$ is a linear model. $Y = x^A$ is *nonlinear*.

LEAST ABSOLUTE DEVIATION REGRESSION

The two most popular linear regression methods for estimating model coefficients are OLS and LAD goodness of fit, respectively. Because they are popular, a wide variety of computer software is available to help us do the calculations.

With *least-squares* goodness of fit, we seek to minimize the sum

$$\sum_i (Y_i - a - bX_i)^2$$

where Y_i denotes the variable we wish to predict and X_i the corresponding value of the predictor on the ith occasion. With the LAD method, we seek to minimize the sum of the absolute deviations between the observed and the predicted value

$$\sum_i |Y_i - a - bX_i|$$

Those who have taken calculus know that the OLS minimum is obtained when

$$\sum_i (Y_i - a - bX_i)b = 0 \quad \text{and} \quad \sum_i (Y_i - a - bX_i) = 0$$

that is, when

$$b = \frac{\text{Covariance}(RM)}{\text{Variance}(R)} = \frac{\sum (R_i - \bar{R})(M_i - \bar{M})}{\sum (R_i - \bar{R})^2}$$

and

$$a = \bar{M} - b\bar{R}$$

LAD attempts to correct one of the major flaws of OLS, that of giving sometimes excessive weight to extreme values. The LAD method solves for those values of the coefficients in the regression equation for which the sum of the absolute deviations $\sum |y_i - R[x_i]|$ is a minimum.

Finding the LAD minimum is more complicated and requires linear programming. But as there is plenty of commercially available software to do the calculations for us, we needn't worry about its complexity.

Algorithms for LAD regression are given in Barrodale and Roberts [1973]. The qreg function of Stata provides for LAD regression, as does R's quantreg package.

LAD regression should be used in preference to OLS in four circumstances:

1. To reduce the influence of outliers.
2. If the losses associated with errors in prediction are additive, rather than large errors being substantially more important than small ones.
3. If the conditional distribution of $Y|X = x$ is not symmetric and we wish to estimate the median of $Y|X = x$ rather than its mean value.
4. If the conditional distribution of $Y|X = x$ is heavy in the tails.

Figure 12.1 depicts systolic blood pressure as a function of age. Each circle corresponds to a pair of observations on a single individual. The solid line is the LAD regression line. The dotted line is the OLS regression line. A single individual, a 47-year-old with a systolic blood pressure of 220, is responsible for the difference between the two lines. Which line do you believe it would be better to use for prediction purposes?

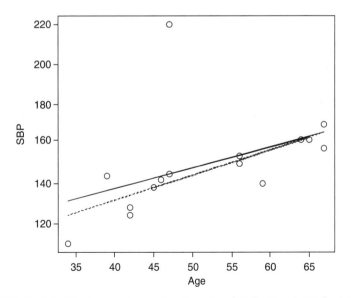

Figure 12.1. Systolic blood pressure as a function of age. Solid line: LAD fit; dotted line: OLS fit.

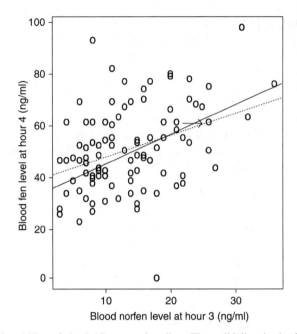

Figure 12.2. Instability of the LAD regression line. The solid line is the LAD fit to the data shown. The dashed line is the LAD line for a perturbed data set that differs from the original solely by a change in the value of one observation by 1/20,000th of the interquartile range of the *x*s. The arrow in the figure indicates the observation that is perturbed and the direction of its perturbation. *Source*: Ellis [1998]. Reprinted with permission from Institute of Mathematical Statistics.

Drawbacks of LAD Regression

On the downside, LAD is unstable in the sense that a small change in the data can cause a relatively large change in the fitted plane. Consider Figure 12.2, which depicts the blood fenuramine level at hour 4 versus the blood norfenuramine level at hour 3 for a group of 99 research subjects in a fenuramine challenge experiment of Malone et al. [1996]. For a perturbed data set that differs from the original solely by a change in the value of one observation by a mere 1/20,000th of the interquartile range of the *x*s. The arrow in the figure indicates the observation that is perturbed and the direction of its perturbation.

ERRORS-IN-VARIABLES REGRESSION

The need for errors-in-variables (EIV) or Deming regression is best illustrated by the struggles of a small medical device firm to bring its product to market. First, they must convince regulators that their long-lasting device provides results equivalent to those of a less efficient device already on the market. In other words, they need to show that the values V recorded by their device bear a linear relation to the values W recorded by their competitor, that is, that $E(V) = a + bW$.

But the errors inherent in measuring W (the so-called predictor) are as large as if not larger than the variation inherent in the output V of the new device. The EIV regression method used to demonstrate equivalence differs in two respects from that of OLS:

1. With OLS, we are trying to minimize the sum of squares $\sum(y_{oi} - y_{pi})^2$, where y_{oi} is the ith observed value of Y and y_{pi} is the ith predicted value. With EIV, we are trying to minimize the sums of squares of errors, going both ways: $\sum(y_{oi} - y_{pi})^2/\text{Var } Y + \sum(x_{oi} - x_{pi})^2/\text{Var } X$.
2. The coefficients of the EIV regression line depend on the ratio $\lambda = \text{Var } X/\text{Var } Y$.

Unfortunately, in cases involving only single measurements by each method, the ratio λ may be unknown and is often assigned a default value of 1. In a simulated

---

WHEN DOES THIS DIFFERENCE MATTER?

When the relative errors for the two methods are similar and the correlation coefficient is greater than 0.8, the OLS regression slope can be approximated as

$$\rho = \text{OLS slope}/\text{Deming slope}$$

where ρ is the correlation coefficient. This means that the regular slope routinely underestimates the actual slope of the data. For ρ less than 0.8, the relationship is no longer as accurate. However differences of 20% or more continue to exist between the slopes calculated by the two methods.

For many clinical chemistry procedures ρ is greater than 0.995, and there is very little difference between OLS and Deming regression. However, for analytes such as electrolytes and many hematology parameters (especially the white cell counts), ρ can easily be less than 0.95 and sometimes in the range of 0.2 to 0.8. In these cases, the use of Deming statistics makes a large difference in the results.

One such example, depicted in Figure 12.3, arises when activated partial thromboplastin time (APTT) is used to determine the correct dose of heparin (a blood thinner). Either too much or too little heparin could seriously impair a patient's health. But which line are we to use?

---

comparison of two electrolyte methods, Linnet [1998] found that misspecification of λ produced a bias that amounted to two-thirds of the maximum bias of the OLS regression method. Standard errors and the results of hypothesis testing also became misleading. In a simulated comparison of two glucose testing methods, Linnet found that a misspecified error ratio resulted in only negligible bias. Given a short range of values in relation to the measurement errors, it is important that λ is correctly estimated, either from duplicate sets of measurements or, in the case of single measurement sets, specified from quality-control data. Even with a misspecified error ratio, Linnet found that Deming regression analysis is likely to perform better than OLS regression analysis.

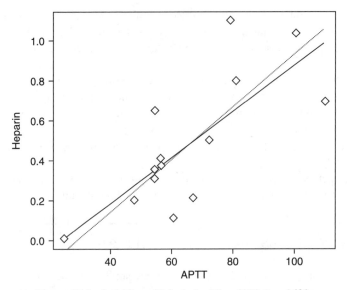

Figure 12.3. Solid line: OLS; dashed line: EIV. $\lambda = 1600$.

In practice, Stöckl, Dewitte, and Thienpont [1998] find that it is not the statistical model but the quality of the analytical input data that is crucial for interpretation of method comparison studies.

Correlation versus Slope of the Regression Line

Perfect correlation ($\rho^2 = 1$) does not imply that two variables are identical but rather that one of them, Y, say, can be written as a linear function of the other, $Y = a + bX$, where b is the slope of the regression line and a is the intercept.

How Big Should the Sample Be?

In method comparison studies, we need to be sure that differences of medical importance are detected. As discussed in Chapter 2, for a given difference, the necessary number of samples depends on the range of values and the analytical standard deviations of the methods involved.

Linnet [1999] finds that the sample sizes of 40–100 conventionally used in method comparison studies often are inadequate. A main factor is the range of values, which should be as wide as possible for the given analyte. For a range ratio (maximum value divided by minimum value) of 2, 544 samples are required to detect one standardized slope deviation; the number of required samples decreases to 64 at a range ratio of 10 (proportional analytical error). For electrolytes having very narrow ranges of values, very large sample sizes usually are necessary. In case of proportional analytical error, application of a weighted approach is important to ensure an efficient analysis; for example, for a range ratio of 10, the weighted approach reduces the sample size requirement by more than 50%.

NINE GUIDELINES

1. Use statistics to provide estimates of errors, not as indicators of acceptability.
2. Recognize that the main purpose of the method comparison experiment is to obtain an estimate of systematic error or bias.
3. Obtain estimates of systematic error at important medical decision concentrations.
4. When there is a single medical decision concentration, make the estimate of systematic error near the mean of the data.
5. When there are two or more medical decision concentrations, use the correlation coefficient, r, to assess whether the range of data is adequate for using ordinary regression analysis.
6. When the correlation coefficient exceeds 0.975, use the comparison plot along with ordinary linear regression statistics.
7. When the correlation coefficient is close to zero, improve the data or change the statistical technique.
8. When in doubt about the validity of the statistical technique, see whether the choice of statistics changes the outcome or decision on acceptability.
9. Plan the experiment carefully and collect the data appropriate for the statistical technique to be used.

Source: Abstracted from Westgard [1998].

QUANTILE REGRESSION

Linear regression techniques (OLS, LAD, or EIV) are designed to help us predict expected values, as in $E(Y) = \mu + \beta X$. But what if our real interest is in predicting extreme values—if, for example, we would like to characterize the observations of Y that are likely to lie in the upper and lower tails of Y's distribution. This would certainly be the case for economists and welfare workers who want to predict the number of individuals whose incomes will place them below the poverty line; physicians, bacteriologists, and public health officers who want to estimate the proportion of bacteria that will remain untouched by various doses of an antibiotic; ecologists and nature lovers who want to estimate the number of species that might perish in a toxic waste spill; and industrialists and retailers who want to know what proportion of the population might be interested in and can afford their new product.

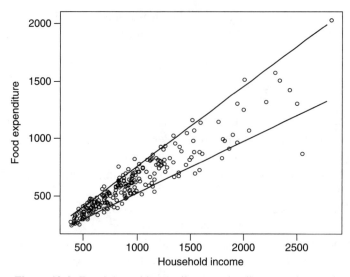

Figure 12.4. Engel data with quantile regression lines superimposed.

In estimating the τth quantile,[1] we try to find that value of β for which $\sum_k \rho_\tau(y_k - f[x_k, \beta])$ is a minimum, where

$$\rho_\tau[x] = \tau x \qquad \text{if } x > 0$$
$$= (\tau - 1)x \quad \text{if } x \le 0$$

Even when expected values or medians lie along a straight line, other quantiles may follow a curved path. Koenker and Hallock [2001] applied the method of quantile regression to data taken from Ernst Engel's 1857 study of the dependence of household food expenditure on household income. As Figure 12.4 reveals, not only was an increase in food expenditures observed as expected when household income lose, but the dispersion of the expenditures also increased.

Some precautions are necessary. As Brian Cade notes, the most common errors associated with quantile regression include:

1. Failing to evaluate whether the model form is appropriate—for example, forcing a linear fit through an obviously nonlinear response. (Of course, this is also a concern with mean regression, OLS, LAD, and EIV.)
2. Trying to overinterpret a single quantile estimate (say, 0.85) with a statistically significant nonzero slope ($p < 0.05$) when the majority of adjacent quantiles (say, 0.5–0.84 and 0.86–0.95) are clearly zero ($p > 0.20$).
3. Failing to use all the information a quantile regression provides. Even if you think you are interested only in relations near the maximum (say, 0.90–0.99),

---

[1] τ is pronounced "tau."

your understanding will be enhanced by having estimates (and sampling variation via confidence intervals) across a wide range of quantiles (say, 0.01–0.99).

THE ECOLOGICAL FALLACY

The court wrote in *NAACP v. City of Niagara Falls*, "Simple regression does not allow for the effects of racial differences in voter turnout; it assumes that turnout rates between racial groups are the same."[2] Whenever distinct strata exist, one should develop separate regression models for each stratum. Failure to do so constitutes the ecological fallacy.

In the 2004 election for governor of the State of Washington, out of the over 2.8 million votes counted, just 261 votes separated the two leading candidates, Christine Gregoire and Dino Rossi, with Mr. Rossi in the lead. Two recounts later, Ms. Gregoire was found to be ahead by 129 votes. There were many problems with the balloting, including the discovery that some 647 felons had voted despite having lost the right to vote. *Borders et al. v. King County et al.* represents an attempt to overturn the results, arguing that if the illegal votes were deducted from each precinct in proportion to the relative number of votes cast for each candidate, Mr. Rossi would have won the election.

> The Court finds that the method of proportionate deduction and the assumption relied upon by Professors Gill and Katz expert witnesses for the plaintiff are a scientifically unaccepted use of the method of ecological inference. In particular, Professors Gill and Katz committed what is referred to as the ecological fallacy in making inferences about a particular individual's voting behavior using only information about the average behavior of groups; in this case, voters assigned to a particular precinct. The ecological fallacy leads to erroneous and misleading results. Election results vary significantly from one similar precinct to another, from one election to another in the same precinct and among different candidates of the same party in the same precinct. Felons and others who vote illegally are not necessarily the same as others in the precinct.
>
> . . . The Court finds that the statistical methods used in the reports of Professors Gill and Katz ignore other significant factors in determining how a person is likely to vote. In this case, in light of the candidates, gender may be as significant or a more significant factor than others. The illegal voters were disproportionately male and less likely to have voted for the female candidate.[3]

To see how stratified regression would be applied in practice, consider a suit[4] to compel redistricting to create a majority Hispanic district in Los Angeles County. The plaintiffs offered in evidence two regression equations to demonstrate differences

---

[2]65 F.3d 1002, n2 (2nd Cir. 1994).

[3]Quotations are from a transcript of the decision by Chelan County Superior Court Judge John Bridges, June 6, 2005.

[4]*Garza et al. v. County of Los Angeles*, 918 F.2d 763 (9th Cir), cert. denied.

in the voting behavior of Hispanics and non-Hispanics:

$$Y_{hi} = C_h + b_h X_{hi} + \varepsilon_{hi}$$
$$Y_{ti} = C_t + b_t X_{hi} + \varepsilon_{ti}$$

where Y_{hi}, Y_{ti} are the predicted proportions of voters in the ith precinct for the Hispanic candidate and for all candidates, respectively; C_h, C_t are the percentages of non-Hispanic voters who voted for the Hispanic candidate and for any candidate; b_h, b_t are the added percentages of Hispanic voters who voted for the Hispanic candidate and for any candidate; X_{hi} is the percentage of registered voters in the ith precinct who are Hispanic; and ε_{hi}, ε_{ti} are random or otherwise unexplained fluctuations.

If there were no differences in the voting behavior of Hispanics and non-Hispanics, then we would expect our estimates of b_h, b_t to be close to zero. Instead, the plaintiffs showed that the best fit to the data was provided by the equations

$$Y_h = 7.4\% + .110 X_h$$
$$Y_t = 42.5\% - .048 X_h$$

Of course, other estimates of the Cs and bs are possible, as only the Xs and Ys are known with certainty; it is conceivable, though unlikely, that few if any of the Hispanics actually voted for the Hispanic candidate.

NONSENSE REGRESSION

Nonlinear regression methods are appropriate when the form of the nonlinear model is known in advance. For example, a typical pharmacological model will have the form $A \exp[bX] + C \exp[dW]$. The presence of numerous locally optimal but globally suboptimal solutions creates challenges, and validation is essential. See, for example, Gallant [1987] and Carroll et al. [1995].

To be avoided are a recent spate of proprietary algorithms available solely in software form which guarantee to find a best-fitting solution. In the words of John von Neumann, "With four parameters I can fit an elephant and with five I can make him wiggle his trunk." Goodness of fit is no guarantee of predictive success, a topic we take up repeatedly in subsequent chapters.

SUMMARY

In this chapter, we distinguished linear from linear regression and described a number of alternatives to ordinary least squares regression, including least absolute deviation regression, and quantile regression. We also noted the importance of using separate regression equations for each identifiable stratum.

TO LEARN MORE

Consider using LAD regression when analyzing software data sets [Miyazaki et al., 1994] or meteorological data [Mielke et al., 1996]. But heed the caveats noted by Ellis [1998].

Only iteratively reweighed general Deming regression produces statistically unbiased estimates of systematic bias and reliable confidence intervals of bias. For details of the recommended technique, see Martin [2000].

Mielke and Berry [2001, Section 5.4] provide a comparison of MRPP, Cade–Richards, and OLS regression methods. Stöckl, Dewitte, and Thienpont [1998] compare ordinary linear regression, Deming regression, standardized principal component analysis, and Passing–Bablok regression.

For more on quantile regression, download Blossom and its accompanying manual from http://www.fort.usgs.gov/products/software/blossom/.

For R code to implement any of the preceding techniques, see Good [2005].

13

MULTIVARIABLE REGRESSION

CAVEATS

Multivariable regression is plagued by the same problems as univariate regression, plus many more of its own. Is the model correct? Are the associations spurious?

In the univariate case, if the errors were not normally distributed, we could take advantage of permutation methods to obtain exact significance levels in tests of the coefficients. Exact permutation methods do not exist in the multivariable case.

When selecting variables to incorporate in a multivariable model, we are forced to perform repeated tests of hypotheses, so that the resultant p-values are no longer meaningful. One solution, if sufficient data are available, is to divide the data set into two parts, using the first part to select variables and the second part to test these variables for significance.

If choosing the correct functional form of a model in a univariate case presents difficulties, consider that in the case of k variables, there are k linear terms (should we use logarithms? should we add polynomial terms?) and $k(k-1)$ first-order cross products of the form $x_i x_k$. Should we include any of the $k(k-1)(k-2)$ second-order cross products?

Should we use forward stepwise regression, or backward regression, or some other method for selecting variables for inclusion? The order of selection can result in major differences in the final form of the model [see, e.g., Roy, 1958; Goldberger, 1961].

Common Errors in Statistics (and How to Avoid Them), Third Edition. Edited by P. I. Good and J. W. Hardin
Copyright © 2009 John Wiley & Sons, Inc.

David Freedman [1983] searched for and found a large and highly significant R^2 among *totally independent* normally distributed random variables. What led him to such an experiment in the first place? How could he possibly have guessed at the results he would obtain?

The Freedman article demonstrates how testing multiple hypotheses, a process that typifies the method of stepwise regression, can only exacerbate the effects of spurious correlation. As he notes in the introduction to the article, "If the number of variables is comparable to the number of data points, and if the variables are only imperfectly correlated among themselves, then a very modest search procedure will produce an equation with a relatively small number of explanatory variables, most of which come in with significant coefficients, and a highly significant R^2. Such an equation is produced even if Y is *totally unrelated* to the Xs."

Freedman used computer simulation to generate 5100 independent, normally distributed random "observations." He put these observations into a data matrix in the format required by the SAS regression procedure. His organization of the values defined 100 observations on each of 51 random variables. Arbitrarily, the first 50 variables were designated as "explanatory" and the 51st as the dependent variable Y.

In the first of two passes through the "data," all 50 explanatory variables were used. Out of the 50 coefficients, 15 were significant at the 25% level, and 1 out of the 50 was significant at the 5% level.

Focusing attention on the explanatory variables that proved significant on the first pass, Freedman constructed a second model using only those variables. The resulting model had an R^2 of 0.36, and the model coefficients of six of the explanatory (but completely unrelated) variables were significant at the 5% level. Given these findings, how can we be sure if the statistically significant variables we uncover in our own research regression models are truly explanatory or are merely the result of chance?

Gail Gong [1986] was among the first, if not the first, student to have the bootstrap as the basis of her doctoral dissertation. Reading her article, we learn that the bootstrap can be an invaluable tool for model validation, a result we explore at greater length in the following chapter. We also learn not to take for granted the results of a stepwise regression.

Gong [1986] constructed a logistic regression model based on observations Peter Gregory made on 155 chronic hepatitis patients, 33 of whom died. The purpose of the model was to identify patients at high risk. In contrast to the computer simulations David Freedman performed, the 19 explanatory variables were real, not simulated, derived from medical histories, physical examinations, X-rays, liver function tests, and biopsies.

If one or more extreme values can influence the slope and intercept of a univariate regression line, think how much more impact, and how subtle the effect, these values might have on a curve drawn through 20-dimensional space.[1]

---

[1]That's 1 dimension for risk of death, the dependent variable, and 19 for the explanatory variables.

Gong's logistic regression models were constructed in two stages. At the first stage, each of the explanatory variables was evaluated on a univariate basis. Thirteen of these variables proved significant at the 5% level when applied to the original data. A forward multiple regression was applied to these 13 variables, and 4 were selected for use in the predictor equation.

When Gong took bootstrap samples of the 155 patients, the R^2 values of the final models associated with each bootstrap sample varied widely. Not reported in this article, but far more important, is that while two of the original four predictor variables always appeared in the final model derived from a bootstrap sample of the patients, five other variables were incorporated in only *some* of the models.

We strongly urge you to adopt Dr. Gong's bootstrap approach to validating multi-variable models. Retain only those variables which appear consistently in the boot-strap regression models. Additional methods for model validation are described in Chapter 15.

CORRECTING FOR CONFOUNDING VARIABLES

When your objective is to verify the association between predetermined explanatory variables and the response variable, multiple linear regression analysis permits you to provide for one or more confounding variables that could not be controlled otherwise.

KEEP IT SIMPLE

It is always best to keep things simple; fools rush in where angels fear to tread. Multivariate regression should be attempted only for exploratory purposes (in the hope of learning more from fewer observations) or as the final stage in a series of modeling attempts of increasing complexity.

Following the 2004 U.S. presidential election, critics noted that (1) the final tabu-lated results differed sharply from those of earlier exit polls and (2) many of the more flagrant discrepancies occurred when ballots were recorded electronically rather than via a paper ballot [Loo, 2005].

The straightforward way to prove or disprove that the electronically recorded ballots were tampered with would be to compare the discrepancies between the exit polls and the final tabulations of precincts which recorded ballots solely by electronic means with the discrepancies observed in a matched set of case controls selected from precincts where paper ballots were used. Surprisingly, Hout et al. [2005] chose instead to build a multivariate regression model in which the depen-dent variable was the 2004 final count for George Bush, and the independent vari-ables included the 2000 final count for Bush, the square of this count, the 1996 final count for Robert Dole, the change in voter turnout, the median income, the proportion of the population that was Hispanic, and whether or not electronic voting machines were used.

DYNAMIC MODELS

Dynamic models are the basis of weather forecasts, long-range models of climate change, and galactic movement. According to Nielsen–Gammon [2003], the following are the chief sources of error in dynamic models:

1. Measurement errors. These tend to be larger at the extremes of each variable.
2. Nonrepresentative measurements (which may result when measurements are taken too far apart in time or in space).
3. Attempting to interpolate between grid points. Here is one example: Suppose that after a particularly strong cold front, there's a strong wind from the north across Texas with cloudy skies and very cold temperatures, say 30°F. As the cold air is blown across the Gulf of Mexico, it gets heated by the warm Gulf waters. So, a grid point 25 km onshore would have a temperature of 30°F and a grid point 25 km offshore might have a temperature of 46°F. Interpolating the model output to the coastline, halfway between the two grid points, gives a temperature of 38°F. But until the air passes over the warm water, it won't start heating up. So, the air will stay at 30°F all the way to the coastline. Simply using interpolated model output (38°F) would have given an 8°F error.

To improve a model:

- Don't merely copy computer output; temper it with your other knowledge of the phenomena you are modeling.
- Refine the model on the basis of the errors observed when it is applied to a test data set. Note that errors may be either of position (in space or in time) or of magnitude.

FACTOR ANALYSIS

The procedures that are involved in factor analysis (FA) as used by psychologists today have several features in common with the procedures for administering Rorschach inkblots. In both procedures, data are first gathered objectively and in quantity; subsequently, the data are analysed according to rational criteria that are time-honoured while not fully understood. . . .

—Chris Brand

Alas, the ad hoc nature of factor analysis is such that one cannot perform the analysis without displeasing somebody. For example, while one group of researchers might argue that a majority of variables should end up identified principally with just one factor, an equally vociferous opposition considers it folly to break up clear g factors by an obsessional search for simple structure.

A factor analysis should be given the same scrutiny as any other modeling procedure and validated as described in Chapter 15. Godino, Batanero, and Jaimez [2001] note that the following errors are frequently associated with factor analysis:

- Applying it to data sets with too few cases in relation to the number of variables analyzed (fewer than two cases per variable in a thesis) without noticing that correlation coefficients have very wide confidence intervals in small samples.
- Using oblique rotation to get a number of factors bigger or smaller than the number of factors obtained in the initial extraction by principal components as a way to show the validity of a questionnaire—for example, obtaining only one factor by principal components and using the oblique rotation to justify that there were two differentiated factors, even when the two factors were correlated and the variance explained by the second factor was very small.
- Confusion among the total variance explained by a factor and the variance explained in the reduced factorial space. In this way, a researcher interpreted that a given group of factors explaining 70% of the variance before rotation could explain 100% of the variance after rotation.

Godino, Batanero, and Jaimez [2001] write, "It is symptomatic that these errors appear in [the work of] doctoral students with a high mathematical preparation, who previously studied analytical geometry. The relevance of the context in the understanding of concepts is shown in these examples. None of these researchers would doubt that a rotation of a solid in the space preserves the solid form (number of factors) and relative dimension of each axis (contribution to the explained variance)."

REPORTING YOUR RESULTS

In reporting the results of your modeling efforts, you need to be explicit about the methods used, the assumptions made, the limitations on your model's range of application, potential sources of bias, and the method of validation (see the following chapter). The section on "Limitations of the Logistic Regression" from Bent and Archfield [2002] is ideal in this regard:

The logistic regression equation developed is applicable for stream sites with drainage areas between 0.02 and 7.00 mi^2 in the South Coastal Basin and between 0.14 and 8.94 mi^2 in the remainder of Massachusetts, because these were the smallest and largest drainage areas used in equation development for their respective areas. [The authors go on to subdivide the area.]

The equation may not be reliable for losing reaches of streams, such as for streams that flow off area underlain by till or bedrock onto an area underlain by stratified-drift deposits (these areas are likely more prevalent where hillsides meet river valleys in central and western Massachusetts). At this juncture of the different underlying surficial deposit types, the stream can lose stream flow through its streambed. Generally, a losing stream reach occurs where the water table does not intersect the streambed in the channel (water table is below the streambed) during low-flow periods. In these reaches, the equation would tend to overestimate the probability of a stream flowing perennially at a site.

The logistic regression equation may not be reliable in areas of Massachusetts where ground-water and surface-water drainage areas for a stream site differ. [The authors go on to provide examples of such areas.]

In these areas, ground water can flow from one basin into another; therefore, in basins that have a larger ground-water contributing area than the surface-water drainage area the equation may underestimate the probability that [a] stream is perennial. Conversely, in areas where the ground-water contributing area is less than the surface-water drainage area, the equation may overestimate the probability that a stream is perennial.

This report also illustrates how data quality, selection, and measurement bias can restrict a model's applicability.

The accuracy of the logistic regression equation is a function of the quality of the data used in its development. This data includes the measured perennial or intermittent status of a stream site, the occurrence of unknown regulation above a site, and the measured basin characteristics.

The measured perennial or intermittent status of stream sites in Massachusetts is based on information in the USGS NWIS database. Stream-flow measured as less than 0.005 ft^3/s is rounded down to zero, so it is possible that several streamflow measurements reported as zero may have had flows less than 0.005 ft^3/s in the stream. This measurement would cause stream sites to be classified as intermittent when they actually are perennial.

Additionally, of the stream sites selected from the NWIS database, 61 of 62 intermittent-stream sites and 89 of 89 perennial-stream sites were represented as perennial streams on USGS topographic maps; therefore, the Statewide database (sample) used in development of the equation may not be random, because stream sites often selected for streamflow measurements are represented as perennial streams on USGS topographic maps. Also, the drainage area of stream sites selected for streamflow measurements generally is greater than about 1.0 mi^2, which may result in the sample not being random.

The observed perennial or intermittent status of stream sites in the South Coastal Basin database may also be biased, because the sites were measured during the summer of 1999. The summer of 1999 did not meet the definition of an extended drought; but monthly precipitation near the South Coastal Basin was less than 50 percent of average in April, less than 25 percent of average in June, about 75 percent of average in July (excluding one station), and about 50 percent of average in August (excluding one station). Additionally, Socolow and others [2000] reported streamflows and ground-water levels well below normal throughout most of Massachusetts during the summer of 1999. Consequently, stream sites classified as intermittent would have been omitted from the database had this period been classified as an extended drought. This climatic condition during the summer of 1999 could bias the logistic regression equation toward a lower probability of a stream site being considered perennial in the South Coastal Basin.

Basin characteristics of the stream sites used in the logistic equation development are limited by the accuracy of the digital data layers used. In the future, digital data layers (such as hydrography, surficial geology, soils, DEMs, and land use) will be at lower scales, such as 1 : 5,000 or 1 : 25,000. This would improve the accuracy of the measured basin characteristics used as explanatory variables to predict the probability of a stream flowing perennially.

For this study, the area of stratified-drift deposits and consequently the areal percentage of stratified-drift deposits included areas with sand and gravel, large sand, fine-grained, and floodplain alluvium deposits. Future studies would allow more specificity in testing the areal percentage of surficial deposits as explanatory variables. For example, the areal percentage of sand and gravel deposits may be an important explanatory variable for estimating the probability that a stream site is perennial. The accuracy of the logistic regression equation also may be improved with the testing of additional basin characteristics as explanatory variables. These explanatory variables could include areal percentage of wetlands (forested and non-forested), areal percentage of water bodies, areal percentage of forested land, areal percentage of urban land, or mean, minimum, and maximum basin elevation.

A CONJECTURE

A great deal of publicity has heralded the arrival of new and more powerful data mining methods, among them neural networks along with dozens of unspecified proprietary algorithms. In our limited experience, none of these have lived up to expectations; see a report of our tribulations in Good [2001, Section 7.6]. Most of the experts we've consulted have attributed this failure to the small size of our test data set, 400 observations each with 30 variables. In fact, many publishers of data mining software assert that their wares are designed solely for use with terrabytes of information

This observation has led to our putting our experience in the form of the following conjecture:

If m points are required to determine a univariate regression line with sufficient precision, then it will take at least m^n observations and perhaps $n!m^n$ observations to appropriately characterize and evaluate a regression model with n variables.

DECISION TREES

As the number of potential predictors increases, the method of linear regression becomes less and less practical. With three potential predictors, we can have as many as seven coefficients to be estimated: one for the intercept, three for first-order terms in the predictors P_i, two for second-order terms of the form $P_i P_j$, and one third-order term $P_1 P_2 P_3$. With k variables, we have k first-order terms, $k(k-1)$ second-order terms, and so forth. Should all these terms be included in our model? Which ones should be neglected? With so many possible combinations, will a single equation be sufficient?

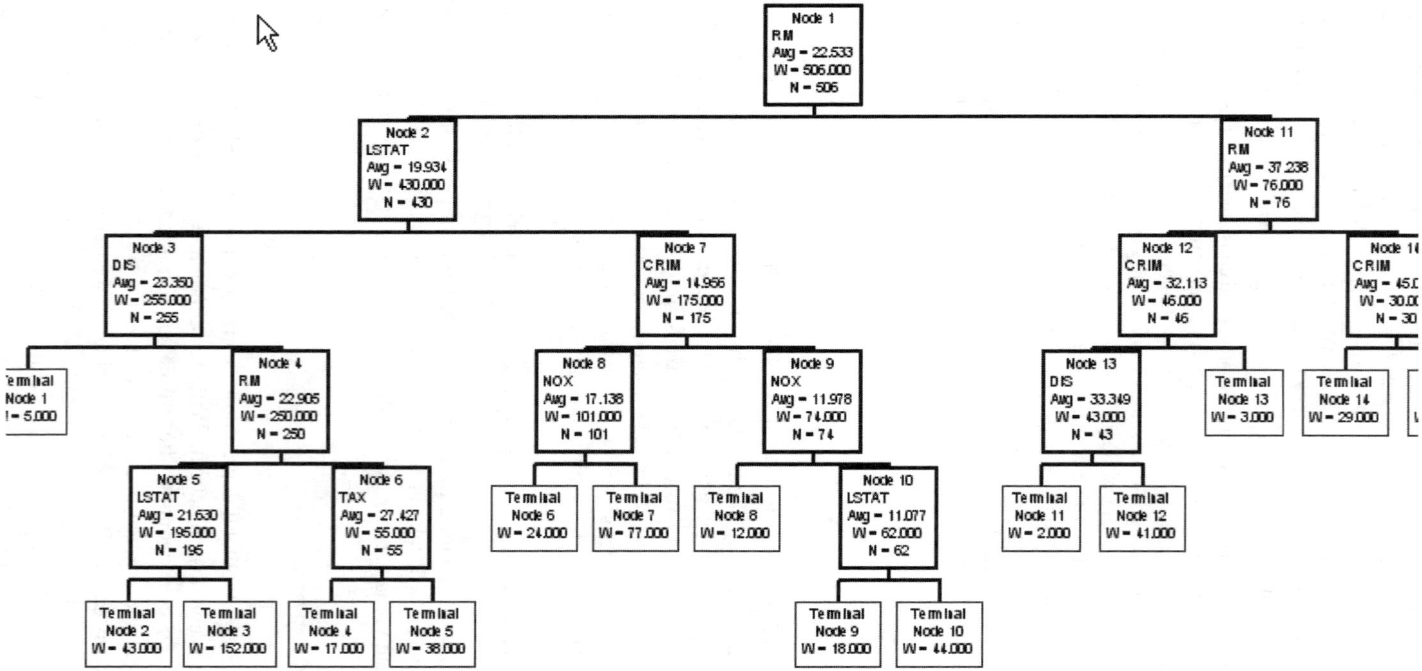

Figure 13.1. CART decision tree for predicting the median house price (OLS).

We need to consider alternate approaches. If you're a mycologist, a botanist, a herpetologist or simply a nature lover, you may have used some sort of key—for example:

1. Leaves simple?
   A. Leaves needle-shaped?
      a. Leaves in clusters of two to many?
         i Leaves in clusters of two to five, sheathed, persistent for several years?

Which is to say that one classifies objects according to whether they possess a particular characteristic. One could accomplish the same result by means of logistic regression, but the latter seems somewhat contrived.

The Classification And Regression Tree (CART) proposed by Breiman, Friedman, Olshen, and Stone [1984] is simply a method of automating the process of classification, so that the initial bifurcation, "Leaves simple" in the preceding example, provides the most effective division of the original sample, and so on. CART can also be used for regression as well as classification, as depicted in Figure 13.1.

We have found CART useful both as a preliminary to multiple regression, as its primary splitters should be used as blocking variables, and for the presentation of results in a readily understandable format, as shown in Table 13.1.

CART offers two other advantages over multiple regression:

1. If known, we can weight the result in terms of the proportions of the various categories in the population at large rather than those in the sample.
2. We can assign losses or penalties on a one-by-one basis to each specific type of misclassification, rather than use some potentially misleading aggregate measure such as least-square error.

Unfortunately, many users fail to take advantage of these latter two features. As always with statistical software, one ought check the defaults to see if they are appropriate for the application at hand.

Alas, regression trees are also known for their instability [Breiman, 1996]. A small change in the training set (see Chapter 15) may lead to a different choice when building a node, which in turn may represent a dramatic change in the tree, particularly if the change occurs in top-level nodes. Branching is also affected by

TABLE 13.1. Comparing Classification (C) and Prediction (R) Methods

Method	OLS	LAD	CART
Estimates	EX = AX	MdnX = Ax	R: either
Loss Function	OLS	LAD	R: OLS, LAD
			C: Arbitrary
Residuals	Symmetric ~ Normal	Symmetric ~ Normal	Arbitrary
Prior Knowledge	N/A	N/A	USe

data density and sparseness, with more branching and smaller bins in data regions where data points are dense. Moreover, in contrast to the smoothness of an OLS regression curve, the jagged approximation provided by CART has marked discontinuities.

REPEATED OBSERVATIONS

Be wary when developing models based on repeated observations on individuals. If your software is permitted to do its own random partitioning, you can get wildly optimistic performance results.

To avoid this, assign the individual, not the record, to a partition so that all records belonging to that individual are either all "train" or all "test."

For quantitative prediction, both regression methods and decision trees have problems. Unthinking use of either approach results in overfitting. With decision trees, this translates into branching rules that seem arbitrary and unrelated to any theory of causation among the variables. Again, this may be the result of a failure to verify the default values.

For example, with the sample data used to predict low birth weight that Salford Systems includes with their product, a decision is based on whether or not the number of first-trimester physician visits (FTV) is two, three, or six. This bizarre finding results from treating FTV as a categorical variable when it is a continuous one.

The complexity of decision trees can be compensated for in part by developing the tree for one set of data and then cross-validating it on another, as described in the next chapter.

BUILDING A SUCCESSFUL MODEL

"Rome was not built in one day,"[2] nor was any reliable model. The only successful approach to modeling involves a continuous cycle of hypothesis formulation, data gathering, hypothesis testing, and estimation. How you navigate through this cycle will depend on whether you are new to the field, have a small data set in hand and are willing and prepared to gather more until the job is done, or have access to databases containing hundreds of thousands of observations. The following prescription, while directly applicable to the latter case, can be readily modified to fit any situation.

1. A thorough literature search and an understanding of causal mechanisms is an essential prerequisite to any study. Don't let the software do your thinking for you.

---

[2]John Heywood, *Proverbes. Part i., Chap. xi., 16th Century.*

2. Using a subset of the data selected at random, see which variables *appear* to be correlated with the dependent variable(s) of interest. (As noted in this and the preceding chapter, two unrelated variables may appear to be correlated by chance alone or as a result of confounding factors. For the same reasons, two closely related factors may fail to exhibit a statistically significant correlation.)

3. Use CART as a preliminary to regression when several categorical variables are involved. Early splits based on the values of categorical variables may suggest that multiple models need to be developed, one for each block. For example, in deciding whether to purchase an item or how many items to purchase, women may make use of different information as well as giving commonly employed information different weights.

4. Using a second distinct subset of the data selected at random, see which of the variables selected at the first stage still *appear* to be correlated with the dependent variable(s) of interest. Alternately, use the bootstrap method describe by Gong [1986] to see which variables are consistently selected for inclusion in the model.

5. Limit attention to one or two of the most significant predictor variables. Select a subset of the existing data in which the remainder of the significant variables are (almost) constant. (Alternately, gather additional data for which the remainder of the significant variables are almost constant.) Decide on a generalized linear model form which best fits your knowledge of the causal relations among the few variables on which you are now focusing. (A standard multivariate linear regression may be viewed as just another form, albeit a particularly straightforward one, of generalized linear model.) Fit this model to the data.

6. Select a second subset of the existing data (or gather an additional data set) for which the remainder of the significant variables are (almost) equal to a second constant. For example, if only men were considered at stage 4, then you should focus on women at this stage. Attempt to fit the model you derived at the preceding stage to these data.

7. By comparing the results obtained at stages 4 and 5, you can determine whether to continue to ignore or to include variables previously excluded from the model. Only one or two additional variables should be added to the model at each iteration of stages 4 through 6.

8. Always validate your results as described in the next chapter.

If all this sounds like a lot of work, it is. It will take several years to develop sound models even or despite the availability of lightning fast, multifunction statistical software. The most common error in statistics is to assume that statistical procedures can take the place of sustained effort.

TO LEARN MORE

Paretz [1981] reviews the effect of autocorrelation on multivariable regression.

Inflation of R^2 as a consequence of multiple tests also is considered by Rencher [1980].

Osborne and Waters [2002] review tests of the assumptions of multivariable regression. Harrell, Lee, and Mark [1996] review the effect of violation of assumptions on generalized linear models and suggest the use of the bootstrap for model validation. Hosmer and Lemeshow [2001] recommend the use of the bootstrap or some other validation procedure before accepting the results of a logistic regression.

Diagnostic procedures for use in determining an appropriate functional form are described by Tukey and Mosteller [1977], Therneau and Grambsch [2000], Hosmer and Lemeshow [2001], and Hardin and Hilbe [2003].

Automated construction of a decision tree dates back to Morgan and Sonquist [1963]. Comparisons of the regression and tree approaches are made by Nurminen [2003] and Perlich, Provost, and Simonoff [2003].

14

MODELING CORRELATED DATA

> While inexact models may mislead, attempting to allow for every contingency a priori is impractical. Thus models must be built by an iterative feedback process in which an initial parsimonious model may be modified when diagnostic checks applied to residuals indicate the need.
>
> —G. E. P. Box

Today, statistical software incorporates advanced algorithms for the analysis of generalized linear models (GLMs)[1] and extensions to panel data settings including fixed-, random-, and mixed-effects models, logistic, Poisson, and negative-binomial regression, generalized estimating equation models (GEEs), and hierarchical linear models (HLMs). These models take the form

$$Y = g^{-1}[\beta X] + \varepsilon,$$

where β is a vector of to-be-determined coefficients, X is a matrix of explanatory variables, and ε is a vector of identically distributed random variables. These variables may follow the normal, gamma, Poisson, or some other distribution, depending on the specified *variance function* of the GLM. The nature of the relationship between the outcome variable and the coefficients depends on the specified *link function* g() of the GLM.

In this chapter, we review popular approaches for modeling correlated data and discuss model properties, assumptions, and relative strengths. We discuss the efficiency gained through correct specification of correlation and the use of alternative standard errors for regression parameters for more robust inference.

---

[1] As first defined by Nelder and Wedderburn [1972].

Common Errors in Statistics (and How to Avoid Them), Third Edition. Edited by P. I. Good and J. W. Hardin
Copyright © 2009 John Wiley & Sons, Inc.

COMMON SOURCES OF ERROR

The caveats of previous chapters also apply to the specification of the link and variance functions of GLMs. The pair of functions which define a specific model should be determined on the basis of cause-and-effect relationships, not by inspecting the data.

For example, when deciding among a Poisson, negative binomial, or binomial model for counts, the wrong approach to model specification is to make function choices based on the ratio of the mean to the variance of the sample. As Bruce Tabor notes in a personal communication:

> In a contagious process, such as an infectious disease outbreak, the probability of a subsequent event will increase after the occurrence of a preceding event. A person carrying an infection is likely to infect additional persons. This results in positive correlation between events and overdispersion—a negative binomial model has this property and may provide a suitable model (or may not, as the case may be).

> In a count process with negative contagion (underdispersion), the occurrence of an event makes subsequent events less likely—events are negatively correlated. One example might be house burglaries in a neighborhood. After an initial burglary, residents and police are alerted to subsequent burglaries and thieves respond appropriately—targeting other neighborhoods for a while.

The other common sources of error in applying GLMs are the use of an inappropriate or erroneous link function, the wrong choice of scale for an explanatory variable (e.g., using x rather than $\log[x]$), neglecting important variables, and the use of an inappropriate error distribution when computing confidence intervals and p-values. Firth [1991, pp. 74–77] should be consulted for a more detailed analysis of potential problems.

PANEL DATA

When multiple observations are collected for each principal sampling unit, we refer to the collected information as panel data, correlated data, or repeated measures. For example, we may collect information on the likelihood that banks offer certain types of loans. If we collect that information from the same set of banks in multiple instances over time, we should expect that observations from the same bank might be correlated.

The dependency of observations violates one of the tenets of regression analysis: that observations are supposed to be i.i.d. Several concerns arise when observations are not independent. First, the effective number of observations (i.e., the effective amount of information) is less than the physical number of observations since, by definition, groups of observations represent the same information. Second, any model that fails to specifically address correlation is incorrect, which means that statistics and tests based on likelihood are based on a faulty specification. Third, while the correct specification of the correlation will yield the most *efficient* estimator, that specification is not the only one to yield a *consistent* estimator.

FIXED- AND RANDOM-EFFECTS MODELS

Most textbooks introduce fixed- and random-effects analysis of variance models through a series of examples. Cases are presented wherein multiple observations are collected for each farm animal or each farm. The basic issue in deciding whether to utilize a fixed- or random-effects model is whether the sampling units (for which multiple observations are collected) represent the collection of most or all of the entities for which inference will be drawn. If so, the fixed-effects estimator is to be preferred. On the other hand, if those same sampling units represent a random sample from a larger population for which we wish to make inferences, then the random-effects estimator is more appropriate.

Fixed- and random-effects models address unobserved heterogeneity. The *random-effects model* assumes that the panel-level effects are randomly distributed. The *fixed-effects model* assumes a constant disturbance that is a special case of the random-effects model. If the random-effects assumption is correct, then the random-effects estimator is more efficient than the fixed-effects estimator. If the random-effects assumption does not hold (i.e., if we specify the wrong distribution for the random effects), then the random-effects model is not consistent. To help decide whether the fixed- or random-effects models is more appropriate, use the Durbin–Wu–Hausman[2] test comparing coefficients from each model.

The fixed-effects approach is sometimes referred to as an assumption-free method since there are no assumptions about the distribution of heterogeneity between the panels. In a meta-analysis combining results from different trials, we might analyze results assuming either fixed or random effects. However, the random-effects assumption may have no medical relevance. In particular, it may not be realistic to assume that the trials combined into our analysis represent some random sample from an underlying population of possible trials. Moreover, there could be selective factors that differ between trials as well as different therapeutic outcomes. Thus, whereas fixed-effects methods may actually be assumption-free, random-effects methods may assume representativeness that is unreasonable. It is often easier to justify application of fixed-effects methods, especially when we focus on the less stringent set of assumptions on which the methods depend.

POPULATION-AVERAGED GEEs

Zeger and Liang [1986] describe a class of estimators that address correlated panel data. The user must specify both a GLM specification valid for independent data and the correlation structure of the panel data.

---

[2]Durbin [1954], Wu [1973], and Hausman [1978] independently discuss this test.

While fixed-effects estimators and random-effects estimators are referred to as subject-specific estimators, the GEEs available through PROC GENMOD in SAS or xtgee in Stata are called population-averaged estimators. This label refer to the interpretation of the fitted regression coefficients. Subject-specific estimators are interpreted in terms of an effect for a given panel, while population-averaged estimators are interpreted in terms of an effect averaged over panels. When and whether to draw inference for average sampling units is considered in the next section.

> The average human has one breast and one testicle.
>
> —Dos McHale

Subject-Specific or Population-Averaged?

A favorite example in comparing subject-specific and population-averaged estimators is to consider the difference in interpretation of regression coefficients for a binary outcome model on whether a child will exhibit symptoms of respiratory illness. The predictor of interest is whether or not the child's mother smokes. Thus, we have repeated observations on children and their mothers. If we were to fit a subject-specific model, we would interpret the coefficient on smoking as the change in the likelihood of respiratory illness as a result of the mother switching from not smoking to smoking.

On the other hand, the interpretation of the coefficient in a population-averaged model is the likelihood of respiratory illness for the average child with a nonsmoking mother compared to the likelihood for the average child with a smoking mother. Both models offer equally valid interpretations. The interpretation of interest should drive model selection; some studies ultimately will lead to fitting both types of models.

> An approximate answer to the right question is worth a good deal more than the exact answer to an approximate problem.
>
> —John W. Tukey

Variance Estimation

In addition to model-based variance estimators, fixed-effects models and GEEs admit *modified sandwich variance estimators*. SAS calls this the empirical variance estimator. Stata refers to it as the "Robust Cluster estimator." Whatever the name, the most desirable property of the variance estimator is that it yields inference for the regression coefficients that is robust to misspecification of the correlation structure.

GEEs require specification of the correlation structure, but the modified sandwich variance estimator (from which confidence intervals and test statistics are constructed) admits inference about the coefficients that is robust to misspecification of that correlation structure. Why then bother with a specification at all? The independence model is an attractive alternative to interpretation of regression coefficients within the more complicated dependence model. Why not then just assume that the observations are independent but utilize this variance estimator in case the independence assumption is incorrect? This is not a recommended approach because the correct specification yields an estimator that is much more efficient than the estimator for an incorrect specification. This efficiency is an asymptotic property of the estimator dependent on the number of independent panels. Zeger and Liang [1986] demonstrate the advantages of correct specification of the correlation structures for GEEs.

Specification of GEEs should include careful consideration of a reasonable correlation structure so that the resulting estimator is as efficient as possible. To protect against misspecification of the correlation structure, one should base inference on the modified sandwich variance estimator. This is the default estimator in SAS, but the user must specify it in Stata. Check your software documentation to ensure best practices.

This same variance estimator is available for the fixed-effects estimator but not for the random-effects estimator.

QUICK REFERENCE FOR POPULAR PANEL ESTIMATORS

Fixed Effects. An indicator variable for each panel/subject is added and used to fit the model. Though often applied to the analysis of repeated measures, this approach has bias that increases with the number of subjects. If data include a very large number of subjects, the associated bias of the results can make this a very poor model choice.

Conditional Fixed Effects. These are commonly applied in logistic regression, Poisson regression, and negative binomial regression. A sufficient statistic for the subject effect is used to derive a conditional likelihood such that the subject-level effect is removed from the estimation.

While conditioning out the subject-level effect in this manner is algebraically attractive, interpretation of model results must continue to be in terms of the conditional likelihood. This may be difficult, and the analyst must be willing to alter the original scientific questions of interest to questions in terms of the conditional likelihood.

Questions always arise as to whether some function of the independent variable might be more appropriate to use than the independent variable itself. For example, suppose $X = Z^2$, where $E(Y \mid Z)$ satisfies the logistic equation; then $E(Y \mid X)$ does not.

Random Effects. The choice of a distribution for the random effect is driven too often by the need to find an analytic solution to the problem rather than by any scientific foundation. If we assume a normally distributed random effect when the random effect is really Laplace, we will obtain the same point estimates (since both distributions are symmetric with mean zero), but we will compute different standard errors. We will have no way of comparing the assumed distributions short of fitting both models.

If the true random-effects distribution has a nonzero mean, then the misspecification is more troublesome, as the point estimates of the fitted model are different from those that would be obtained from fitting the true model. Knowledge of the true random-effects distribution does not alter the interpretation of fitted model results. Instead, we are limited to discussing the relationship of the fitted parameters to those parameters we would obtain if we had access to the entire population of subjects, and we fit that population to the same fitted model. In other words, even given the knowledge of the true random-effects distribution, we cannot easily compare fitted results to true parameters.

As discussed in Chapter 6 with respect to group-randomized trials, if the subjects are not independent (suppose that they all come from the same classroom), then the true random effect is actually larger. The attenuation of our fitted coefficient increases as a function of the number of supergroups containing our subjects as members; if classrooms are within schools and there is within-school correlation, the attenuation is even greater.

Compared to fixed-effects models, random-effects models have the advantage of using up fewer degrees of freedom, but they have the disadvantage of requiring that the regressors be uncorrelated with the disturbances; this last requirement should be checked with the Durbin–Wu–Hausman test.

GEE. Instead of trying to derive the estimating equation for the GLM with correlated observations from a likelihood argument, the within-subject correlation is introduced directly into the estimating equation of an independence model. The correlation parameters are then nuisance parameters and can be estimated separately. [See also Hardin and Hilbe, 2003.]

Underlying the population-averaged GEE is the assumption that one is able to specify the correct correlation structure. If one hypothesizes an exchangeable correlation and the true correlation is time dependent, the resulting regression coefficient estimator is inefficient. The naive variance estimates of the regression coefficients will then produce incorrect confidence intervals. Analysts specify a correlation structure to gain efficiency in the estimation of the regression coefficients, but typically they calculate the sandwich estimate of variance to protect against misspecification of the correlation. This variance estimator is more variable than the naive variance estimator, and many analysts do not pay adequate attention to the fact that the asymptotic properties depend on the number of subjects (not the total number of observations).

HLM. This includes hierarchical linear models, linear latent models, and others. While previous models are limited for the most part to a single effect, HLM allows more than one. Unfortunately, most commercially available software requires one to assume that each random effect is Gaussian with mean zero. The variance of each random effect must be estimated. As we cautioned in the section on random effects, the choice of distribution should be carefully investigated. Litière, Alonso, and Mohlenberghs [2008] discuss the impact of misspecifying the random-effect distributions on inferential procedures.

Mixed Models. These models allow both linear and nonlinear mixed effects regression (with various links). They allow you to specify each level of repeated measures (imagine: districts: schools: teachers: classes: students). In this description, each of the sublevels is within the previous level, and we can hypothesize a fixed or random effect for each level. We also imagine that observations within the same levels (any of these specific levels) are correlated.

TO LEARN MORE

For more on the contrast between fixed-effect assumption-free methods and random-effect assumed-representativeness methods, see Section 5.17 of http://www.ctsu.ox.ac.uk/reports/ebctcg-1990/section5.

See Hardin and Hilbe [2003, p. 28] for a more detailed explanation of specifying the correlation structure in population-averaged GEEs. See Zeger and Liang [1998] for detailed investigations of efficiency and consistency for misspecified correlation structures in population-averaged GEEs.

See McCullagh and Nelder [1989] and Hardin and Hilbe [2007] for the theory and application of GLMs. See Skrondal and Rabe-Hesketh [2004] for extensions of GLMs to include latent variables and to structural equation models. For more information on longitudinal data analysis utilizing specific software, Stata users should see Rabe-Hesketh and Skrondal [2008], and SAS users should see Cody [2001].

15

VALIDATION

> "... the simple idea of splitting a sample in two and then developing the hypothesis on the basis of one part and testing it on the remainder may perhaps be said to be one of the most seriously neglected ideas in statistics, if we measure the degree of neglect by the ratio of the number of cases where a method could give help to the number of cases where it is actually used."
> —G. A. Barnard in discussion following Stone [1974, p. 133]

Validate your models before drawing conclusions.

Absent a detailed knowledge of causal mechanisms, the results of a regression analysis are highly suspect. Freedman [1983] found highly significant correlations between totally independent variables. Gong [1986] resampled repeatedly from the data in hand and obtained a different set of significant variables each time.

OBJECTIVES

A host of advertisements for new proprietary software claim an ability to uncover relationships previously hidden and to overcome the deficiencies of linear regression. But how can we determine whether or not such claims are true?

Good [2001, Chapter 10] reports on one such claim from the maker of PolyAnalyst™. He took the 400 records, each of 31 variables, that PolyAnalyst provided in an example data set, split the data in half at random, and obtained completely discordant results with the two halves whether they were analyzed with PolyAnalyst, CART, or stepwise linear regression. This is yet another example of a spurious relationship that did not survive the validation process.

Common Errors in Statistics (and How to Avoid Them), Third Edition. Edited by P. I. Good and J. W. Hardin
Copyright © 2009 John Wiley & Sons, Inc.

In this chapter, we review the various methods of validation and provide guidelines for their application.

METHODS OF VALIDATION

Your choice of an appropriate methodology will depend upon your objectives and the stage of your investigation. Is the purpose of your model to predict (will there be an epidemic?), to extrapolate (What might the climate have been like on the primitive Earth?), or to elicit causal mechanisms (is development accelerating or decelerating?) (Which factors are responsible?)?

Are you still developing the model and selecting variables for inclusion or are you in the process of estimating model coefficients?

There are three main approaches to validation:

1. Independent verification (obtained by waiting until the future arrives or through the use of surrogate variables).
2. Splitting the sample (using one part for calibration, the other for verification). See Picaro and Cook [1984].
3. Resampling (taking repeated samples from the original sample and refitting the model each time).

Independent Verification

Independent verification is appropriate and preferable whatever the objectives of your model and whether selecting variables for inclusion or estimating model coefficients.

In soil, geologic, and economic studies, researchers often return to the original setting and take samples from points that were bypassed on the original round. See, for example, Tsai et al. [2001].

In other studies, verification of the model's form and the choice of variables are obtained by attempting to fit the same model in a similar but distinct context.

For example, having successfully predicted an epidemic at one army base, one would then wish to see if a similar model might be applied at a second and a third almost but not quite identical base.

Stockton and Meko [1983] reconstructed regional-average precipitation to A.D. 1700 in the Great Plains of the United States with multiple linear regression models calibrated on the period 1933–1977. They validated the reconstruction by comparing the reconstructed regional percentage-of-normal precipitation with single-station precipitation for stations with records extending back as far as the 1870s. Lack of an appreciable drop in correlation between these single-station records and the reconstruction from the calibration period to the earlier segment was taken as evidence for validation of the reconstructions.

Graumlich [1993] used a response-surface reconstruction method to reconstruct 1000 years of temperature and precipitation in the Sierra Nevada. The calibration climatic data were 62 years of observed precipitation and temperature (1928–1989) at Giant Forest/Grant Grove. The model was validated by comparing the predictions with the 1873–1927 segments of three climate stations 90 km to the west in the San Joaquin Valley. The climatic records of these stations were highly correlated with those at Giant Forest/Grant Grove. Significant correlation of these long-term station records with the 1873–1927 part of the reconstruction was accepted as evidence of validation.

Independent verification can help discriminate among several models that appear to provide equally good fits to the data. Independent verification can be used in conjunction with either of the two other validation methods. For example, an automobile manufacturer was trying to forecast parts sales. After correcting for seasonal effects and long-term growth within each region, Auto Regressive Integrated Moving Average (ARIMA) techniques were used.[1] A series of best-fitting ARIMA models was derived, one model for each of the nine sales regions into which the sales territory had been divided. The nine models were quite different in nature. As the regional seasonal effects and long-term growth trends had been removed, a single ARIMA model applicable to all regions, albeit with differing coefficients, was more plausible. Accordingly, the ARIMA model that gave the best overall fit to all regions was utilized for prediction purposes.

Independent verification also can be obtained through the use of surrogate or proxy variables. For example, we may want to investigate past climates and test a model of the evolution of a regional or worldwide climate over time. We cannot go back directly to a period before direct measurements on temperature and rainfall were made, but we can observe the width of growth rings in long-lived trees or measure the amount of carbon dioxide in ice cores.

Sample Splitting

Splitting the sample into two parts, one for estimating the model parameters, the other for verification, is particularly appropriate for validating time series models where the emphasis is on prediction or reconstruction. If the observations form a time series, the more recent observations should be reserved for validation purposes. Otherwise, the data used for validation should be drawn at random from the entire sample.

Unfortunately, when we split the sample and use only a portion of it, the resulting estimates will be less precise.

Browne [1975] suggests we pool rather than split the sample if

1. The predictor variables to be employed are specified beforehand (i.e., we do not use the information in the sample to select them).
2. The coefficient estimates obtained from a calibration sample drawn from a certain population are to be applied to other members of the same population.

---

[1] For examples and discussion of ARIMA processes, see Brockwell and Davis [1987].

The proportion to be set aside for validation purposes will depend upon the loss function. If both the goodness-of-fit error in the calibration sample and the prediction error in the validation sample are based on mean squared error, Picard and Berk [1990] report that we can minimize their sum by using between a quarter and a third of the sample for validation purposes.

A compromise proposed by Moiser [1951] is worth revisiting: The original sample is split in half; regression variables and coefficients are selected independently for each of the subsamples; if they are more or less in agreement, then the two samples should be combined and the coefficients recalculated with greater precision.

A further proposal by Subrahmanyam [1972] to use weighted averages where there are differences strikes us as equivalent to painting over cracks left by the last earthquake. Such differences are a signal to probe deeper, to look into causal mechanisms, and to isolate influential observations which may, for reasons which need to be explored, be marching to a different drummer.

Resampling

We saw in the report of Gail Gong [1985], discussed in Chapter 13, that resampling methods such as the bootstrap may be used to validate our choice of variables to include in the model. As seen in the previous chapter, they may also be used to estimate the precision of our estimates.

But if we are to extrapolate successfully from our original sample to the population at large, then our original sample must bear a strong resemblance to that population. When only a single predictor variable is involved, a sample of 25 to 100 observations may suffice. But when we work with n variables simultaneously, sample sizes on the order of 25^n to 100^n may be required to adequately represent the full n-dimensional region.

Because of dependencies among the predictors, we can probably get by with several orders of magnitude fewer data points. But the fact remains that the sample size required for confidence in our validated predictions grows exponentially with the number of variables.

Five resampling techniques are in general use:

1. *K-fold*, in which we subdivide the data into K roughly equal-sized parts and then repeat the modeling process K times, leaving one section out each time for validation purposes.

2. *Leave-one-out*, an extreme example of K-fold, in which we subdivide the data into as many parts as there are observations. We leave one observation out of our classification procedure and use the remaining $n - 1$ observations as a training set. Repeating this procedure n times, omitting a different observation each time, we arrive at a figure for the number and percentage of observations classified correctly. A method that requires this much computation would have been unthinkable before the advent of inexpensive, readily available, high-speed computers. Today, at worst, we need step out for a cup of coffee while our desktop computer completes its work.

3. *Jackknife*, an obvious generalization of the leave-one-out approach, where the number left out can range from one observation to half of the sample.

4. *Delete-d*, where we set aside a random percentage d of the observations for validation purposes, use the remaining $100 - d\%$ as a training set, then average over 100 to 200 such independent random samples.

5. The *bootstrap*, which we have already considered at length in earlier chapters.

The correct choice among these methods in any given instance is still a matter of controversy (though any statistician will assure you that the matter is quite settled). See, for example, Wu [1986] and the discussion following and Shao and Tu [1995].

Leave-one-out has the advantage of allowing us to study the influence of specific observations on the overall outcome.

Our own opinion is that if any of the above methods suggest that the model is unstable, the first step is to redefine the model over a more restricted range of the various variables. For example, with the data of Figure 11.3, we would advocate confining attention to observations for which the predictor (TNFAlpha) was less than 200.

If a more general model is desired, then many additional observations should be taken in underrepresented ranges. In the cited example, this would be values of TNFAlpha greater than 300.

MEASURES OF PREDICTIVE SUCCESS

Whatever method of validation is used, we need to have some measure of the success of the prediction procedure. One possibility is to use the sum of the losses in the calibration and the validation sample. Even this procedure contains an ambiguity that we need resolve. Are we more concerned with minimizing the expected loss, the average loss, or the maximum loss?

One measure of goodness of fit of the model is SSE $= \sum (y_i - y_i^*)^2$ where y_i and y_i^* denote the ith observed value and the corresponding value obtained from the model. The smaller this sum of squares, the better the fit.

If the observations are independent, then

$$\sum (y_i - y_i^*)^2 = \sum (y_i - \bar{y})^2 - \sum (\bar{y} - y_i^*)^2$$

The first sum on the right-hand side of the equation is the total sum of squares (SST). Most statistics software uses as a measure of fit $R^2 = 1 - \text{SSE}/\text{SST}$. The closer the value of R^2 is to 1, the better.

The automated entry of predictors into the regression equation using R^2 runs the risk of overfitting, as R^2 is guaranteed to increase with each predictor entering the model. To compensate, one may use the adjusted R^2

$$1 - [((n - i)(1 - R^2))/(n - p)]$$

where n is the number of observations used in fitting the model, p is the number of estimated regression coefficients, and i is an indicator variable that is 1 if the model includes an intercept and 0 otherwise.

The adjusted R^2 has two major drawbacks, according to Rencher and Pun [1980]:

1. The adjustment algorithm assumes that the predictors are independent; more often, the predictors are correlated.
2. If the pool of potential predictors is large, multiple tests are performed, and R^2 is inflated in consequence; the standard algorithm for adjusted R^2 does not correct for this inflation.

A preferable method of guarding against overfitting the regression model, proposed by Wilks [1995], is to use validation as a guide for stopping the entry of additional predictors. Overfitting is judged to begin when entry of an additional predictor fails to reduce the prediction error in the validation sample.

Mielke et al. [1997] propose the following measure of predictive accuracy for use with either a mean-square-deviation or a mean-absolute-deviation loss function:

$$M = 1 - \delta/\mu_\delta$$

where

$$\delta = \frac{1}{n} \sum_{i=1}^{n} |y_i - y_i^*| \quad \text{and} \quad \mu_\delta = \frac{1}{n^2} \sum_{i=1}^{n} \sum_{j=1}^{n} |y_i - y_j^*|$$

Uncertainty in Predictions

Whatever measure is used, the degree of uncertainty in your predictions should be reported. Error bars are commonly used for this purpose.

The prediction error is larger when the predictor data are far from their calibration-period means and vice versa. For simple linear regression, the standard error of the estimate s_e and the standard error of prediction s_{y*} are related as follows:

$$s_{y*} = s_e \sqrt{\frac{(n+1)}{n} + (x_p - \bar{x})^2 / \sum_{i=1}^{n} (x_i - \bar{x})^2}$$

where n is the number of observations and x_i is the ith value of the predictor in the calibration sample, and x_p is the value of the predictor used for the prediction.

The relation between s_{y*} and s_e is easily generalized to the multivariate case. In matrix terms, if $Y = AX + E$ and $y^* = AX_p$, then $s_{y*}^2 = s_e^2\{1 + x_p^T(X^TX)^{-1}x_p\}$.

This equation is applicable only if the vector of predictors lies inside the multivariate cluster of observations on which the model was based. An important question is, how different can the predictor data be from the values observed in the calibration period before the predictions are considered invalid?

LONG-TERM STABILITY

Time is a hidden dimension in most economic models. Many an airline has discovered to its detriment that today's optimal price leads to half-filled planes and markedly reduced profits tomorrow. A careful reading of the newspapers lets them know that a competitor has slashed prices, but more advanced algorithms are needed to detect a slow shift in the tastes of prospective passengers. The public, tired of being treated no better than hogs,[2] turns to trains, personal automobiles, and teleconferencing.

An army base, used to a slow seasonal turnover in recruits, suddenly finds that all infirmary beds are occupied and the morning lineup for sick call stretches the length of a barracks.

To avoid a pound of cure:

- Treat every model as tentative, best described, as any lawyer will advise you, as subject to change without notice.
- Monitor continuously.

Most monitoring algorithms take the following form:

If the actual value exceeds some boundary value (the series mean, for example, or the series mean plus one standard deviation).

And if the actual value exceeds the predicted value for three observation periods in a row.

Sound the alarm (if the change, like an epidemic, is expected to be temporary in nature) or recalibrate the model.

TO LEARN MORE

Almost always, a model developed on one set of data will fail to fit a second independent sample nearly as well. Mielke et al. [1996] investigated the effects of sample size, type of regression model, and noise to signal ratio on the decrease or shrinkage in fit from the calibration to the validation data set.

For more on leave-one-out validation see Michaelsen [1987], Weisberg [1985], and Barnston and van den Dool [1993]. Camstra and Boomsma [1992] and Shao and Tu [1995] review the application of resampling in regression.

Miller, Hui, and Tierney [1991] propose validation techniques for logistic regression models.

Watterson [1996] reviews the various measures of predictive accuracy.

---

[2]Or somewhat worse, because hogs generally have a higher percentage of fresh air to breathe.

GLOSSARY, GROUPED BY RELATED BUT DISTINCT TERMS

Accuracy and precision   An *accurate* estimate is close to the estimated quantity. A *precise* interval estimate is a narrow one. Precise measurements made with a dozen or more decimal places may still not be accurate.

Deterministic and stochastic   A phenomenon is *deterministic* when its outcome is inevitable and all observations will take specific value.[1] A phenomenon is *stochastic* when its outcome may take different values in accordance with some probability distribution.

Dichotomous, categorical, ordinal, metric data   *Dichotomous* data have two values and take the form "yes or no," "got better or got worse."

Categorical data have two or more categories such as "yes," "no," and "undecided." Categorical data may be ordered (opposed, indifferent, in favor) or unordered (dichotomous, categorical, ordinal, metric).

Preferences can be placed on an ordered or *ordinal* scale such as "strongly opposed," "opposed," "indifferent," "in favor," "strongly in favor."

Metric data can be placed on a scale that permits meaningful subtraction; for example, while "in favor" minus "indifferent" may not be meaningful, 35.6 pounds minus 30.2 pounds is.

Metric data can be grouped so as to evaluate them by statistical methods applicable to categorical or ordinal data. But to do so would be to throw away information, and would reduce the power of any tests and the precision of any estimates.

---

[1]These observations may be subject to measurement error.

Common Errors in Statistics (and How to Avoid Them), Third Edition.   Edited by P. I. Good and J. W. Hardin
Copyright © 2009 John Wiley & Sons, Inc.

Distribution, cumulative distribution, empirical distribution, limiting distribution Suppose we were able to examine all the items in a population and record a value for each one to obtain a *distribution* of values. The *cumulative distribution function* of the population $F[x]$ denotes the probability that an item selected at random from this population will have a value less than or equal to x. $0 \leq F[x] \leq 1$. Also, if $x < y$, then $F[x] \leq F[y]$.

The *empirical distribution*, usually represented in the form of a cumulative frequency polygon or a bar plot, is the distribution of values observed in a sample taken from a population. If $F_n[x]$ denotes the cumulative distribution of observations in a sample of size n, then as the size of the sample increases, $F_n[x] \rightarrow F[x]$.

The *limiting distribution* for very large samples of a sample statistic such as the mean or the number of events in a large number of very small intervals often tends to a distribution of known form such as the Gaussian for the mean or the Poisson for the number of events.

Be wary of choosing a statistical procedures which is optimal only for a limiting distribution and not when applied to a small sample. For a small sample, the empirical distribution may be a better guide.

Hypothesis, null hypothesis, alternative The dictionary definition of a *hypothesis* is a proposition, or set of propositions, put forth as an explanation for certain phenomena.

For statisticians, a *simple hypothesis* would be that the distribution from which an observation is drawn takes a specific form. For example, $F[x] = N(0, 1)$. In the majority of cases, a statistical hypothesis will be *compound* rather than simple—for example, that the distribution from which an observation is drawn has a mean of zero.

Often, it is more convenient to test a *null hypothesis*—for example, that there is no or null difference between the parameters of two populations.

There is no point in performing an experiment or conducting a survey unless one also has one or more *alternate hypotheses* in mind. If the alternative is one-sided—for example, the difference is positive rather than zero—then the corresponding test will be one-sided. If the alternative is two-sided—for example, the difference is not zero—then the corresponding test will be two-sided.

Parametric, nonparametric, and semiparametric models Models can be subdivided into two components, one systematic and one random. The systematic component can be a function of certain predetermined parameters (a parametric model), be parameter free (nonparametric), or be a mixture of the two types (semiparametric). The definitions in the following section apply to the random component.

Parametric, nonparametric, and semiparametric statistical procedures *Parametric* statistical procedures concern the parameters of distributions of a known form. One may want to estimate the variance of a normal distribution or the number of degrees of freedom of a chi-square distribution. Student's t, the F-ratio, and maximum likelihood are typical parametric procedures.

Nonparametric procedures concern distributions whose form is unspecified. One might use a nonparametric procedure like the bootstrap to obtain an interval estimate for a mean or a median or to test that the distributions of observations drawn from two different populations are the same. Nonparametric procedures are often referred to as distribution-free, though not all distribution-free procedures are nonparametric in nature.

Semiparametric statistical procedures concern the parameters of distributions whose form is not specified. Permutation methods and U-statistics are typically employed in a semiparametric context.

Residuals and errors   A residual is the difference between a fitted value and what was actually observed. An error is the difference between what is predicted based on a model and what is actually observed.

Significance level and p-value   The *significance level* is the probability of making a Type I error. It is a characteristic of a statistical procedure.

The *p-value* is a random variable that depends both upon the sample and upon the statistical procedure that is used to analyze the sample.

If one repeatedly applies a statistical procedure at a specific significance level to distinct samples taken from the same population when the hypothesis is true and all assumptions are satisfied, then the p-value will be less than or equal to the significance level with the frequency given by the significance level.

Type I and Type II error   A Type I error is the probability of rejecting the hypothesis when it is true. A Type II error is the probability of accepting the hypothesis when an alternative hypothesis is true. Thus, a Type II error depends on the alternative.

Type II error and power   The power of a test for a given alternative hypothesis is the probability of rejecting the original hypothesis when the alternative is true. A Type II error is made when the original hypothesis is accepted even though the alternative is true. Thus, power is 1 minus the probability of making a Type II error.

BIBLIOGRAPHY

Adams DC; Gurevitch J; Rosenberg MS. Resampling tests for meta-analysis of ecological data. *Ecology* 1997; 78:1277–1283.

Aickin M; Gensler H. Adjusting for multiple testing when reporting research results: The Bonferroni vs Holm methods. *Am. J. Publ. Health* 1996; 85:726–728.

Altman DG. Statistics in medical journals. *Statist. Med.* 1982; 1:59–71.

Altman DG. Randomisation. *BMJ.* 1991; 302:1481–1482.

Altman DG. The scandal of poor medical research. *BMJ* 1994; 308:283–284.

Altman DG. Statistics in medical journals: Developments in the 1980s. *Statist. Med.* 1991; 10:1897–1913.

Altman DG. Statistical reviewing for medical journals. *Statist. Med.* 1998; 17:2662–2674.

Altman DG. Commentary: Within trial variation—A false trail? *J. Clin. Epidemiol.* 1998; 51:301–303.

Altman DG. Statistics in medical journals: Some recent trends. *Statist. Med.* 2000; 19:3275–3289.

Altman DG. Poor quality medical research: What can journals do? *JAMA* 2002.

Altman DG; De Stavola BL; Love SB; Stepniewska KA. Review of survival analyses published in cancer journals. *Br. J. Cancer* 1995; 72:511–518.

Altman DG; Lausen B; Sauerbrei W; Schumacher M. Dangers of using "optimal" cutpoints in the evaluation of prognostic factors. [Commentary] *JNCI* 1994; 86:829–835.

Altman DG; Schulz KF; Moher D; Egger M; Davidoff F; Elbourne D; Gøtzsche PC. Lang T for the CONSORT Group. The revised consort statement for reporting randomized trials: Explanation and elaboration. *Annals Internal Med.* 2001; 134:663–694.

Common Errors in Statistics (and How to Avoid Them), Third Edition. Edited by P. I. Good and J. W. Hardin
Copyright © 2009 John Wiley & Sons, Inc.

Aly E-E AA. Simple test for dispersive ordering. *Statist. Prob. Letters* 1990; 9:323–325.

Andersen B. *Methodological Errors in Medical Research*. Blackwell, Oxford, 1990.

Anderson DR; Burnham KP; Thompson WL. Null hypothesis testing: Problems, prevalence, and an alternative. *J. Wildlife Management* 2000; 64:912–923.

Anderson S; Hauck WW. A proposal for interpreting and reporting negative studies. *Statist. Med.* 1986; 5:203–209.

Anscombe F. Sequential medical trials (book review). *JASA* 1963; 58:365.

Armitage P. Test for linear trend in proportions and frequencies. *Biometrics* 1955; 11:375–386.

Avram MJ; Shanks CA; Dykes MHM; Ronai AK; Stiers WM. Statistical methods in anesthesia articles: An evaluation of two American journals during two six-month periods. *Anesthesia and Analgesia* 1985; 64:607–611.

Bacchetti P. Peer review of statistics in medical research: The other problem. *BMJ* 2002; 324:1271–1273.

Badrick TC; Flatman RJ. The inappropriate use of statistics. *NZ J. Med. Lab. Sci.* 1999; 53:95–103.

Bailar JC; Mosteller F. Guidelines for statistical reporting in articles for medical journals. Amplifications and explanations. *Annals of Internal Medicine* 1988; 108:66–73.

Bailey KR. Inter-study differences: How should they influence the interpretation and analysis of results? *Statist. Med.* 1987; 6:351–358.

Bailor AJ. Testing variance equality with randomization tests. *Statist. Comp. Simul.* 1989; 31:1–8.

Baker RD. Two permutation tests of equality of variance. *Statist. Comput.* 1995; 5(4):289–296.

Balakrishnan N; Ma CW. A comparative study of various tests for the equality of two population variances. *Statist. Comp. Simul.* 1990; 35:41–89.

Barbui C; Violante A; Garattini S. Does placebo help establish equivalence in trials of new antidepressants? *Eur. Psychiatry* 2000; 15:268–273.

Barnston AG; van den Dool HM. A degeneracy in cross-validated skill in regression-based forecasts. *J. Climate* 1993; 6:963–977.

Barrodale I; Roberts FDK. An improved algorithm for discrete l_1 linear approximations. *Soc. Industr. Appl. Math. J. Numerical Anal.* 1973; 10:839–848.

Bayarri MJ; Berger J. Quantifying surprise in the data and model verification. In: Bernado et al., eds. *Bayesian Statistics*. Oxford: Oxford University Press, 1998; 53–82.

Bayes T. An essay toward solving a problem in the doctrine of chances. *Philosophical Transactions of the Royal Society* 1763; 53:370–418.

Begg C; Berlin J. (with discussion). Publication bias: A problem in interpreting medical data. *JRSS A* 1988; 151:419–436.

Begg CB; Cho M; Eastwood S; Horton R; Moher D; Olkin I; Pitkin R; Rennie D; Schulz KF; Simel D; Stroup DF. Improving the quality of reporting of randomized controlled trials: The CONSORT Statement. *JAMA* 1996; 276:637–639.

Bent GC; Archfield SA. A logistic regression equation for estimating the probability of a stream flowing perennially in Massachusetts USGC. Water-Resources Investigations Report 02-4043.

Berger JO. *Statistical Decision Theory and Bayesian Analysis*. 2nd ed. Springer-Verlag: New York, 1986.

Berger JO. Could Fisher, Jefferies, and Neyman have agreed on testing? *Statist. Sci.* 2003; 18:1–32.

Berger JO; Berry DA. Statistical analysis and the illusion of objectivity. *The American Scientist* 1988; 76:159–165.

Berger JO; Selike T. Testing a point null hypothesis: The irreconcilability of *P*-values and evidence. *JASA* 1987; 82:112–122.

Berger VW. Pros and cons of permutation tests. *Statist. Med.* 2000; 19:1319–1328.

Berger VW. Improving the information content of endpoints in clinical trials. *Controlled Clinical Trials* 2002; 23:502–514.

Berger VW. *Selection Bias and Covariate Imbalances in Randomized Clinical Trials*. Wiley, 2005.

Berger VW. Response to Klassen et al. Missing data should be more heartily penalized. *Journal of Clinical Epidemiology* 2006; 59:759–761.

Berger VW, Exner DV. Detecting selection bias in randomized clinical trials. *Controlled Clinical Trials* 1999; 20:319–327.

Berger VW; Lunneborg C; Ernst MD; Levine JG. Parametric analyses in randomized clinical trials. *J. Modern Appl. Statist. Meth.* 2002; 1:74–82.

Berger VW; Ivanova A. Bias of linear rank tests for stochastic order in ordered categorical data. *J. Statist. Planning and Inference* 2002; 107:237–297.

Berger VW; Permutt T; Ivanova A. Convex hull test of ordered categorical data. *Biometrics* 1998; 54:1541–1550.

Berkeley G. *Treatise Concerning the Principles of Human Knowledge*. Oxford University Press, 1710.

Berkey C; Hoaglin D; Mosteller F; Colditz G. A random effects regression model for meta-analysis. *Statist. Med.* 1995; 14:395–411.

Berkson J. Tests of significance considered as evidence. *JASA* 1942; 37:325–335.

Berlin JA; Laird NM; Sacks HS; Chalmers TC. A comparison of statistical methods for combining event rates from clinical trials. *Statist. Med.* 1989; 8:141–151.

Berry DA. Decision analysis and Bayesian methods in clinical trials. In *Recent Advances in Clinical Trial Design and Analysis*. 125–154. Kluwer Press: New York, (Ed: Thall P), 1995.

Berry DA. *Statistics: A Bayesian Perspective*. Duxbury Press: Belmont, California, 1996.

Berry DA; Stangl DK. *Bayesian Biostatistics*. Marcel Dekker: New York, 1996.

Bickel P; Klassen CA; Ritov Y; Wellner S. *Efficient and Adaptive Estimation for Semi-Parametric Models*. Johns Hopkins University Press: Baltimore, 1993.

Bishop G; Talbot M. Statistical thinking for novice researchers in the biological sciences. In: Batanero C. (ed.), *Training Researchers in the Use of Statistics*. International Association for Statistical Education and International Statistical Institute. Granada, Spain, 2001; 215–226.

Bland JM; Altman DG. Comparing methods of measurement: Why plotting difference against standard method is misleading. *Lancet* 1995; 346:1085–1087.

Block G. A review of validations of dietary assessment methods. *Am. J. Epidemiol.* 1982; 115:492–505.

Bly RW. *Power-Packed Direct Mail: How to Get More Leads and Sales by Mail*. Henry Holt, 1996.

Bly RW. *The Copywriter's Handbook: A Step-By-Step Guide to Writing Copy That Sells.* Henry Holt, 1990.

Blyth CR. On the inference and decision models of statistics (with discussion). *Ann. Statist.* 1970; 41:1034–1058.

Bothun G. *Modern Cosmological Observations and Problems.* Taylor and Francis: London, 1998.

Box GEP; Anderson SL. Permutation theory in the development of robust criteria and the study of departures from assumptions. *JRSS-B* 1955; 17:1–34.

Box GEP; Tiao GC. A note on criterion robustness and inference robustness. *Biometrika* 1964; 51:169–173.

Breiman L. Bagging Predictors. *Machine Learning* 1996; 24:123–140.

Breiman L; Friedman JH; Olshen RA; Stone CJ. *Classification and Regression Trees.* Wadsworth and Brooks: Monterey CA, 1984.

Brockwell PJ; Davis RA. *Time Series: Theory and Methods.* Springer-Verlag: New York, 1987.

Browne MW. A comparison of single sample and cross-validation methods for estimating the mean squared error of prediction in multiple linear regression. *British J. Math. Statistist Psychol.* 1975; 28:112–120.

Buchanan-Wollaston H. The philosophic basis of statistical analysis. *J. Int. Council Explor. Sea* 1935; 10:249–263.

Burn DA. Designing effective statistical graphs. In: Rao CR. (ed.), *Handbook of Statistics.* 1993; 9, Chapter 22.

Buyse M; Piedbois P. On the relationship between response to treatment and survival time. *Statistics in Medicine* 1996; 15:2797–2812.

Cade B; Richards L. Permutation tests for least absolute deviation regression. *Biometrics* 1996; 52:886–902.

Callaham ML; Wears RL; Weber EJ; Barton C; Young G. Positive-outcome bias and other limitations in the outcome of research abstracts submitted to a scientific meeting. *JAMA* 1998; 280:254–257.

Camstra A; Boomsma A. Cross-validation in regression and covariance structure analysis. *Sociological Methods and Research* 1992; 21:89–115.

Canty AJ; Davison AC; Hinkley DV; Ventura V. Bootstrap diagnostics. http://www.stat.cmu. edu/www/cmu-stats/tr/tr726/tr726.html.

Cappuccio FP; Elliott P; Allender PS; Pryer J; Follman DA; Cutler JA. Epidemiologic association between dietary calcium intake and blood pressure: A meta-analysis of published data. *Am. J. Epidemiol.* 1995; 142:935–945.

Carleton RA; Lasater TM; Assaf AR; Feldman HA; McKinlay S. et al. The Pawtucket Heart Health Program: Community changes in cardiovascular risk factors and projected disease risk. *Am. J. Public Health* 1995; 85:777–785.

Carlin BP; Louis TA. *Bayes and Empirical Bayes Methods For Data Analysis.* Chapman and Hall: London, UK, 1996.

Carmer SG; Walker WM. Baby bear's dilemma: A statistical tale. *Agronomy Journal* 1982; 74:122–124.

Carpenter J; Bithell J. Bootstrap confidence intervals. *Statist. Med.* 2000; 19:1141–1164.

Carroll RJ; Ruppert D. Transformations in regression: A robust analysis. *Technometrics* 1985; 27:1–12.

Carroll RJ; Ruppert D. *Transformation and Weighting in Regression.* CRC, 2000.

Casella G; Berger RL. *Statistical Inference.* Pacific Grove CA: Wadsworth, Brooks, 1990.

Chalmers TC. Problems induced by meta-analyses. *Statist. Med.* 1991; 10:971–980.

Chalmers TC; Frank CS; Reitman D. Minimizing the three stages of publication bias. *JAMA* 1990; 263:1392–1395.

Charlton BG. The future of clinical research: From megatrials towards methodological rigour and representative sampling. *J. Eval. Clin. Pract.* 1996; 2:159–169.

Chernick MR; Liu CY. The saw-toothed behavior of power versus sample size and software solutions: Single binomial proportion using exact methods. *American Statistician* 2002; 56:149–155.

Cherry S. Statistical tests in publications of The Wildlife Society. *Wildlife Society Bulletin* 1998; 26:947–953.

Chiles JR. *Inviting Disaster: Lessons from the Edge of Technology.* Harper-Collins: New York, 2001.

Choi BCK. Development of indicators for occupational health and safety surveillance. Asian-Pacific Newsletter 2000; 7. http://www.ttl.fi/Internet/English/Information/Electronic+journals/Asian-Pacific+Newsletter/2000-01/04.htm.

Clemen RT. Combining forecasts: A review and annotated bibliography. *International Journal of Forecasting* 1989; 5:559–583.

Clemen RT. *Making Hard Decisions.* PWS–Kent, Boston, 1991.

Clemen RT; Jones SK; Winkler RL. Aggregating forecasts: An empirical evaluation of some Bayesian methods. In *Bayesian Analysis in Statistics and Econometrics.* Berry DA; Chaloner K. (ed). John Wiley and Sons: New York, 1996; 3–13.

Cleveland WS. *The Elements of Graphing Data.* Hobart Press: Summit NJ, 1985.

Cochran WG. *Sampling Techniques.* 3rd ed. New York: Wiley, 1977.

Cody R. *Longitudinal Data And SAS: A Programmer's Guide.* SAS Press, Cary, NC, 2001.

Cohen J. Things I have learned (so far). *American Psychologist* 1990; 45:1304–1312.

Collins R; Keech A; Peto R; Sleight P; Kjekshus J; Wilhelmsen L; et al. Cholesterol and total mortality: Need for larger trials. *BMJ* 1992; 304:1689.

Collins RJ; Weeks JR; Cooper MM; Good PI; Russell RR. Prediction of abuse liability of drugs using intravenous self-administration by rats. *Psychopharmacology* 1984; 82:6–13.

Conover W; Salsburg D. *Biometrics* 1988; 44:189–196.

Conover WJ; Johnson ME; Johnson MM. Comparative study of tests for homogeneity of variances: With applications to the outer continental shelf bidding data. *Technometrics* 1981; 23:351–361.

Converse JM; Presser S. *Survey Questions: Handcrafting the Standardized Questionaire.* Sage, 1986.

Cooper HM; Rosenthal R. Statistical versus traditional procedures for summarising research findings. *Psychol. Bull.* 1980; 87:442–449.

Copas JB; Li HG. Inference for non-random samples (with discussion). *JRSS* 1997; 59:55–95.

Cornfield J; Tukey JW. Average values of mean squares in factorials. *Ann. Math. Statist.* 1956; 27:907–949.

Cox DR. Some problems connected with statistical inference. *Ann. Math. Statist.* 1958; 29:357–372.

Cox DR. The role of significance tests. *Scand J. Statist.* 1977; 4:49–70.

Cox DR. Seven common errors in statistics and causality. *JRSS A* 1992; 155:291.

Cox DR. Some remarks on consulting. *Liaison* (Statistical Society of Canada). 1999; 13:28–30.

Cummings P; Koepsell TD. Statistical and design issues in studies of groups. *Inj. Prev.* 2002; 8:6–7.

Dar R; Serlin; Omer H. Misuse of statistical tests in three decades of psychotherapy research. *J. Consult. Clin. Psychol.* 1994; 62:75–82.

Davision AC; Hinkley DV. *Bootstrap Methods and Their Application*. Cambridge University Press, 1997.

Davision AC; Snell EJ. Residuals and diagnostics. In *Statistical Theory and Modelling*. Hinkley DV; Reid N; Shell EJ. eds. Chapman and Hall: London, 1991; 83.

Day S. Blinding or masking. In *Encyclopedia of Biostatistics*, v1, Armitage P; Colton T. eds. John Wiley and Sons: Chichester, 1998.

Delucchi KL. The use and misuse of chisquare. Lewis and Burke revisited. *Psych. Bull.* 1983; 94:166–176.

Diaconis P. Statistical problems in ESP research. *Science* 1978; 201:131–136.

Diciccio TJ; Romano JP. A review of bootstrap confidence intervals (with discussion). *JRSS B* 1988; 50:338–354.

Dixon PM. Assessing effect and no effect with equivalence tests. In Newman MC; Strojan CL. eds. *Risk Assessment: Logic and Measurement*. Chelsea (MI): Ann Arbor Press, 1998.

Djulbegovic B; Lacevic M; Cantor A; Fields KK; Bennett CL; Adams JR; Kuderer NM; Lyman GH. The uncertainty principle and industry-sponsored research. *Lancet* 2000; 356:635–638.

Donner A; Brown KS; Brasher P. A methodological review of non-therapeutic intervention trials employing cluster randomization, 1979–1989. *Int. J. Epidemiol.* 1990; 19:795–800.

Duggan TJ; Dean CW. Common misinterpretations of significance levels in sociological Journals. *Amer. Sociologist.* 1968; February, 45–46.

Durbin J. Errors in variables. *Revue de l'Institut International de Statistique*. 1954; 22:23–32.

Dyke G. How to avoid bad statistics. *Field Crops Research* 1997; 51:165–197.

Easterbrook PJ; Berlin JA; Gopalan R, Matthews DR. Publication bias in clinical research. *Lancet* 1991; 337:867–872.

Edwards W; Lindman H; Savage L. Bayesian statistical inference for psychological research. *Psychol. Rev.* 1963; 70:193–242.

Efron B. Bootstrap methods, another look at the jackknife. *Annals Statist.* 1979; 7:1–26.

Efron B. *The Jackknife, the Bootstrap, and Other Resampling Plans*. Philadelphia: SIAM, 1982.

Efron B. Better bootstrap confidence intervals, (with discussion). *JASA* 1987; 82:171–200.

Efron B. Bootstrap confidence intervals: Good or bad? (with discussion). *Psychol. Bull.* 1988; 104:293–296.

Efron B. Six questions raised by the bootstrap. R. LePage and L. Billard, eds. *Exploring the Limits of the Bootstrap*. New York: Wiley, 1992.

Efron B; Morris C. Stein's paradox in statistics. *Sci. Amer.* 1977; 236:119–127.

Efron B; Tibshirani R. Bootstrap methods for standard errors, confidence intervals, and other measures of statistical accuracy. *Statist. Sci.* 1986; 1:54–77.

Efron B; Tibshirani R. *An Introduction to the Bootstrap*. New York: Chapman and Hall, 1993.

Egger M; Smith GD. Meta-analysis: Potentials and promise. *BMJ* 1997; 315:1371–1374.

Egger M; Smith GD; Phillips AN. Meta-analysis: Principles and procedures. *BMJ* 1997; 315:1533–1537.

Egger M; Schneider M; Smith GD. Spurious precision? Meta-analysis of observational studies. *British Med. J.* 1998; 316:140–143.

Ehrenberg ASC. Rudiments of numeracy. *JRSS Series A* 1977; 140:277–297.

Ellis SP. Instability of least squares, least absolute deviation and least median of squares linear regression. *Statist. Sci.* 1998; 13:337–350.

Elwood JM. *Critical Appraisal Of Epidemiological Studies And Clinical Trials.* 2nd ed. New York: Oxford University Press, 1998.

Estepa A; Sánchez Cobo FT. Empirical research on the understanding of association and implications for the training of researchers. In: Batanero C. (ed.), *Training Researchers in the Use of Statistics.* International Association for Statistical Education and International Statistical Institute. Granada, Spain, 2001; 37–51.

Eysenbach G; Sa E-R. Code of conduct is needed for publishing raw data. *BMJ* 2001; 323:166.

Farquhar AB; Farquhar H. *Economic and Industrial Delusions: A Discussion of the Case for Protection.* G.P. Putnam's Sons: New York and London. 1851.

Fears TR; Tarone RE; Chu KC. False-positive and false-negative rates for carcinogenicity screens. *Cancer Res.* 1977; 37:1941–1945.

Feinstein AR. P-values and confidence intervals: Two sides of the same unsatisfactory coin. *J. Clin. Epidem.* 1998; 51:355–360.

Feinstein AR; Concato J. The quest for "power": Contradictory hypotheses and inflated sample sizes. *J. Clin. Epidem.* 1998; 51:537–545.

Feller W. *An Introduction to Probability Theory and Its Applications.* v2. Wiley: New York, 1966.

Felson DT; Anderson JJ; Meenan RF. The comparative efficacy and toxicity of second-line drugs in rheumatoid arthritis. *Arthritis and Rheumatism* 1990; 33:1449–1461.

Felson DT; Cupples LA; Meenan RF. Misuse of statistical methods in *Arthritis and Rheumatism.* 1982 versus 1967–1968. *Arthritis and Rheumatism* 1984; 27:1018–1022.

Feng Z; Grizzle J. Correlated binomial variates: Properties of estimator of ICC and its effect on sample size calculation. *Statist. Med.* 1992; 11:1607–1614.

Feng Z; McLerran D; Grizzle J. A comparison of statistical methods for clustered data analysis with Gaussian error. *Statist. Med.* 1996; 15:1793–1806.

Feng Z; Diehr P; Peterson A; McLerran D. Selected statistical issues in group randomized trials. *Annual Rev. Public Health* 2001; 22:167–187.

Fergusson D; Glass KC; Waring D; Shapiro S. Turning a blind eye: The success of blinding reported in a random sample of randomised, placebo controlled trials. *BMJ* 2004; 328:432.

Fienberg SE. Damned lies and statistics: Misrepresentations of honest data. In: Editorial Policy Committee. *Ethics and Policy in Scientific Publications.* Council of Biology Editors, 1990; 202–206.

Fink A; Kosecoff JB. *How to Conduct Surveys: A Step by Step Guide.* Sage, 1988.

Finney DJ. The responsible referee. *Biometrics* 1997; 53:715–719.

Firth D. General linear models. In *Statistical Theory and Modelling.* Hinkley DV; Reid N; Shell EJ. eds. Chapman and Hall: London, 1991; 55.

Fisher NI; Hall P. On bootstrap hypothesis testing. *Australian J. Statist.* 1990; 32:177–190.

Fisher NI; Hall P. Bootstrap algorithms for small samples. *J. Statist. Plan Infer.* 1991; 27:157–169.

Fisher RA. *Design of Experiments.* New York: Hafner, 1935.

Fisher RA. *Statistical Methods and Scientific Inference.* 3rd ed. New York: Macmillan, 1973.

Fleming TR. Surrogate markers in AIDs and cancer trials. *Statist. Med.* 1995; 13:1423–1435.

Fligner MA; Killeen TJ. Distribution-free two-sample tests for scale. *JASA* 1976; 71:210–212.

Fowler FJ jr; Fowler FJ. *Improving Survey Questions: Design and Evaluation.* Sage, 1995.

Frank D; Trzos RJ; Good P. Evaluating drug-induced chromosome alterations. *Mutation Res.* 1978; 56:311–317.

Freedman DA. A note on screening regression equations. *Amer. Statist.* 1983; 37:152–155.

Freedman DA. From association to causation. *Statist. Sci.* 1999.

Freedman DA; Navidi W; Peters SC. On the impact of variable selection in fitting regression equations. In: Dijkstra TK. (ed.), *On Model Uncertainty and Its Statistical Implications.* Springer: Berlin, 1988; 1–16.

Freiman JA; Chalmers TC; Smith H; Kuebler RR. The importance of beta; the type II error; and sample size in the design and interpretation of the randomized controlled trial. In: Bailar JC; Mosteller F; eds. *Medical Uses of Statistics.* Boston: MA, NEJM Books, 1992; 357.

Friedman LM; Furberg CD; DeMets DL. *Fundamentals Of Clinical Trials.* 3rd ed. St. Louis: Mosby, 1996.

Friedman M. The use of ranks to avoid the assumption of normality implicit in the analysis of variance. *JASA* 1937; 32:675–701.

Fritts HC; Guiot J; Gordon GA. Verification. In: Cook ER; Kairiukstis LA. eds. *Methods of Dendrochronology; Applications in the Environmental Sciences.* Kluwer Academic Publishers, 1990; 178–185.

Fujita T; Ohue T; Fuji Y; Miyauchi A; Takagi Y. Effect of calcium supplement on bone density and parathyroid function in elderly subjects. *Miner Electrolyte Metabolism* 1995; 21:229–231.

Fujita T; Ohue T; Fuji Y; Miyauchi A; Takagi Y. Heated oyster shell-seaweed calcium (AAA Ca) on osteoporosis. *Calcified Tissue International* 1996; 58:226–230.

Fujita T; Fujii Y; Goto B; Miyauchi A; Takagi Y. Peripheral computed tomography (pQCT) detected short-term effect of AAACa (heated oyster shell with heated algal ingredient HAI): A double-blind comparison with CaCO3 and placebo. *J Bone Miner Metab.* 2000; 18:212–215.

Fujita T; Ohue T; Fuji Y; Miyauchi A; Takagi Y. Reappraisal of the Katsuragi Calcium study, a prospective, double-blind, placebo-controlled study of the effect of active absorbable algal calcium (AAACa) on vertebral deformity and fracture. *J. Bone Mineral Metabolism* 2004; 22:32–38.

Fukada S. Effects of active amino acid calcium: Its bioavailability in intestinal absorption and removal of plutonium in animals. *J. Bone and Mineral Metabolism* 1993; 11:S47–S51.

Gail MH; Byar DP; Pechacek TF; Corle DK. Aspects of statistical design for the Community Intervention Trial for Smoking Cessation (COMMIT). *Cont. Clin. Trials* 1992; 123:6–21.

Gail MH; Mark SD; Carroll R; Green S; Pee D. On design considerations and randomization-based inference for community intervention trials. *Statist. Med.* 1996; 15:1069–1092.

Gail MH; Tan WY; Piantadosi S. Tests for no treatment effect in randomized clinical trials. *Biometrika.* 1988; 75:57–64.

Gardner MJ; Altman DG. Confidence intervals rather than *P* values: Estimation rather than hypothesis testing. *BMJ* 1996; 292:746–750.

Gardner MJ; Bond J. An exploratory study of statistical assessment of papers published in the *Journal of the American Medical Association*. *JAMA* 1990; 263:1355–1357.

Gardner MJ; Machin D; Campbell MJ. Use of check lists in assessing the statistical content of medical studies. *BMJ* 1986; 292:810–812.

Garthwaite PH. Confidence intervals from randomization tests. *Biometrics* 1996; 52: 1387–1393.

Gastwirth JL; Rubin H. Effect of dependence on the level of some one-sample tests. *JASA* 1971; 66:816–820.

Gavarret J. *Principes Généraux de Statistique Medicale*. Libraires de la Faculte de Medecine de Paris, Paris, 1840.

Geary RC. Testing normality. *Biometrika* 1947; 34:241.

George SL. Statistics in medical journals: A survey of current policies and proposals for editors. *Medical and Pediatric Oncology* 1985; 13:109–112.

Geweke JK; DeGroot MH. *Optimal Statistical Decisions*. McGraw-Hill: New York, 1970.

Gigerenzer G. *Calculated Risks: How To Know When Numbers Deceive You*. Simon; Schuster: NY, 2002.

Gill J. Whose variance is it anyway? Interpreting empirical models with state-level data. *State Politics and Policy Quarterly* 2001; Fall: 318–338.

Gillett R. Meta-analysis and bias in research reviews. *Journal of Reproductive and Infant Psychology* 2001; 19:287–294.

Gine E; Zinn J. Necessary conditions for a bootstrap of the mean. *Ann. Statist.* 1989; 17: 684–691.

Glantz S. Biostatistics: How to detect: correct: and prevent errors in the medical literature. *Circulation* 1980; 61:1–7.

Glass GV; Peckham PD; Sanders JR. Consequences of failure to meet the assumptions underlying the fixed effects analysis of variance and covariance. *Reviews in Educational Research* 1972; 42:237–288.

Godino JD; Batanero C; Gutiérrez-Jaimez RG. The statistical consultancy workshop as a pedagogical tool. In: Batanero C. (ed.), *Training Researchers in the Use of Statistics*. Granada: International Association for Statistical Education and International Statistical Institute. 2001; 339–353.

Goldberger AS. Note on stepwise least squares. *JASA* 1961; 56–293:105–110.

Gong G. Cross-validation, the jackknife and the bootstrap: Excess error in forward logistic regression. *JASA* 1986; 81:108–113.

Gonzales GF; Cordova A; Gonzales C; Chung A; Vega K; Villena A. Lepidium meyenii (Maca) improved semen parameters in adult men. *Asian J. Andrology* 2001; 4:301–303.

Good IJ. *Probability and the Weighing of Evidence*. London: Griffin, 1950.

Good IJ. The Bayes/non-Bayes compromise: A brief review. *JASA*. 1992; 87:597–606.

Good PI. Almost most powerful tests against composite alternatives. *Commun. Statist.* 1989; 18:1913.

Good PI. Most powerful tests for use in matched pair experiments when data may be censored. *J. Statist. Comput. Simul.* 1991; 38:57–63.

Good PI. Globally almost most powerful tests for censored data. *J. Nonpar. Statist.* 1992; 1:253–262.

Good PI. *Applying Statistics in the Courtroom.* Chapman and Hall/CRC, 2001.

Good PI. *Permutation, Parametric, and Bootstrap Tests of Hypotheses.* 3rd ed. New York: Springer, 2005.

Good PI. *Resampling Methods.* 3rd ed. Boston: Birkhauser, 2006.

Good PI. Extensions of the concept of exchangeability and their applications to testing hypotheses. *J. Modern Stat. Anal.* 2002; 2:243–247.

Good PI. *Managers' Guide to the Design and Conduct of Clinical Trials,* 2nd ed. Wiley, NY, 2006.

Good P; Lunneborg CE. Limitations of the analysis of variance. The one-way design. *J. Modern Appl. Statist. Methods* 2005; 5:41–43.

Good P; Xie F. Analysis of a crossover clinical trial by permutation methods. *Contemporary Clinical Trials* 2008; 29:565–568.

Goodman SN. Towards evidence-based medical statistics. II. The Bayes Factor. *Ann. Intern. Med.* 1999; 130:1005–1013.

Goodman SN. Of p-values and Bayes: A modest proposal. *Epidemiology* 2001; 12:295–297.

Goodman SN; Altman DG; George SL. Statistical reviewing policies of medical journals: Caveat lector? *J. Gen. Intern. Med.* 1998; 13:753–756.

Gore S; Jones IG; Rytter EC. Misuse of statistical methods: Critical assessment of articles in BMJ from January to March 1976. *BMJ* 1977; 1:85–87.

Götzsche PC. Reference bias in reports of drug trials. *BMJ* 1987; 295:654–656.

Götzsche PC; Podenphant J; Olesen M; Halberg P. Meta-analysis of second-line antirheumatic drugs: Sample size bias and uncertain benefit. *J. Clin. Epidemiol.* 1992; 45:587–594.

Grant A. Reporting controlled trials. *British J. Obstetrics and GynaEcology* 1989; 96:397–400.

Graumlich L. A 1000-year record of temperature and precipitation in the Sierra Nevada. *Quaternary Research* 1993; 39:249–255.

Green PJ; Silverman BW. *Nonparametric Regression and Generalized Linear Models.* Chapman and Hall: London, 1994.

Greene HL; Roden DM; Katz RJ et al. The Cardiac Arrhythmia Suppression Trial: First CAST . . . then CAST II. *J. Am. Coll. Cardiol.* 1992; 19:894–898.

Greenland S. Modeling and variable selection in epidemiologic analysis. *Am. J. Public. Health* 1989; 79:340–349.

Greenland S. Randomization, statistics, and causal inference. *Epidemiology* 1990; 1:421–429.

Greenland S. Probability logic and probabilistic induction [see comments]. *Epidemiology* 1998; 9:322–332.

Gurevitch J; Hedges LV. Meta-analysis: Combining the results of independent studies in experimental ecology. In: Scheiner S; Gurevitch J. eds. *The Design and Analysis of Ecological Experiments.* Chapman and Hall: London, 1993; 378–398.

Guthery FS; Lusk JJ; Peterson MJ. The fall of the null hypothesis: Liabilities and opportunities. *J. Wildlife Management* 2001; 65:379–384.

Guttorp P. *Stochastic Modeling of Scientific Data.* Chapman and Hall: London, 1995.

Hagood MJ. *Statistics for Sociologists.* NY: Reynal and Hitchcock, 1941.

Hall P; Wilson SR. Two guidelines for bootstrap hypothesis testing. *Biometrics* 1991; 47:757–762.

Hardin JW; Hilbe JM. *Generalized Estimating Equations*. Chapman and Hall/CRC: London, 2003.

Hardin JW; Hilbe JM. *Generalized Linear Models and Extensions*. 2nd ed. Stata Press, College Station, TX, 2007.

Harley SJ; Myers RA. Hierarchical Bayesian models of length-specific catchability of research trawl surveys. *Canadian J. Fisheries Aquatic Sciences* 2001; 58:1569–1584.

Harrell FE; Lee KL. A comparison of the discrimination of discriminant analysis and logistic regression under multivariate normality. In *Biostatistics: Statistics in Biomedical; Public Health; and Environmental Sciences. The Bernard G. Greenberg Volume*. Sen PK. (ed.), New York: North-Holland, 1985; 333–343.

Harrell FE; Lee KL; Mark DB. Multivariable prognostic models: Issues in developing models; evaluating assumptions and adequacy; and measuring and reducing errors. *Statist. Med.* 1996; 15:361–387.

Hastie T; Tibshirani R; Friedman JH. *The Elements of Statistical Learning: Data Mining, Inference, and Prediction*. Springer, 2001.

Hausman JA. Specification tests in econometrics. *Econometrica* 1978; 46:1251–1271.

Hedges LV; Olkin I. *Statistical Methods for Meta-Analysis*. Academic Press: New York, 1985.

Hertwig R; Todd PM. Biases to the left, fallacies to the right: Stuck in the middle with null hypothesis significance testing. *Psycoloquy* 2000; 11: #28. (with discussion).

Hilton J. The appropriateness of the Wilcoxon test in ordinal data. *Statist. Med.* 1996; 15: 631–645.

Hinkley DV; Shi S. Importance sampling and the nested bootstrap. *Biometrika* 1989; 76: 435–446.

Hoenig JM; Heisey DM. The abuse of power: The pervasive fallacy of power calculations for data analysis. *Amer. Statist.* 2001; 55:19–24.

Horowitz RI. Large scale randomised evidence; large simple trials and overviews of trials: Discussion—a clinician's perspective on meta-analysis. *J. Clin. Epidemiol.* 1995; 48: 41–44.

Horwitz RI; Singer BH; Makuch RW; Viscolia CM. Clinical versus statistical considerations in the design and analysis of clinical research. *J. Clinical Epidemiology* 1998; 51:305–307.

Hosmer DW; Lemeshow SL. *Applied Logistic Regression*. Wiley: NY, 2001.

Hout M; Mangels L; Carlson J; Best R. Working paper: The effect of electronic voting machines on change in support for Bush in the 2004 Florida elections. http://www.yuricareport.com/ElectionAftermath04/BerkeleyElection04_WP.pdf. 2005.

Hsu JC. *Multiple Comparisons: Theory and Methods*. Chapman and Hall/CRC, 1996.

Huber PJ. *Robust Statistics* Wiley: New York, 1981.

Hume D. *An Enquiry Concerning Human Understanding*. Oxford University Press, 1748.

Hunter JE; Schmidt FL. Eight common but false objections to the discontinuation of significance testing in the analysis of research data. In: Harlow LL; Mulaik SA; Steiger JH. eds. *What If There Were No Significance Tests?* Lawrence Erlbaum Assoc., Mahwah, NJ, 1997; 37–64.

Hurlbert SH. Pseudoreplication and the design of ecological field experiments. *Ecological Monographs* 1984; 54:198–211.

Husted JA; Cook RJ; Farewell VT; Gladman DD. Methods for assessing responsiveness: A critical review and recommendations. *J. Clinical Epidemiology* 2000; 53:459–468.

Hutchon DJR. Infopoints: Publishing raw data and real time statistical analysis on e-journals. *BMJ* 2001; 322:530.

International Committee of Medical Journal Editors. Uniform requirements for manuscripts submitted to biomedical journals. *JAMA* 1997; 277:927–934.

International Study of Infarct Survival Collaborative Group. Randomized trial of intravenous streptokinase, oral aspirin, both or neither, among 17187 cases of suspected acute myocardial infarction. ISIS-2. *Lancet* 1988; 2:349–362.

Jagers P. Invariance in the linear model—an argument for chi-square and f in nonnormal situations. *Mathematische Operationsforschung und Statistik* 1980; 11:455–464.

Jennison C; Turnbull BW. *Group Sequential Methods with Applications to Clinical Trials*. CRC, 1999.

Johnson DH. The insignificance of statistical significance testing. *J. Wildlife Management* 1999; 63:763–772.

Jones LV. Statistics and research design. *Annual Review Psych.* 1955; 6:405–430.

Jones LV; Tukey JW. A sensible formulation of the significance test. *Psychol. Meth.* 2000; 5:411–416.

Kadane IB; Dickey J; Winklcr R; Smith W; Peters S. Interactive elicitation of opinion for a normal linear model. *JASA* 1980; 75:845–854.

Kanarek MS; Conforti PM; Jackson LA; Cooper RC; Murchio JC. Asbestos in drinking water and cancer incidence in the San Francisco Bay Area. *Amer. J. Epidemiol.* 1980; 112:54–72.

Kaplan J. Misuses of statistics in the study of intelligence: The case of Arthur Jensen (with disc). *Chance* 2001; 14:14–26.

Kass R; Raftery A. Bayes factors. *JASA* 1995; 90:773–795.

Kelly E; Campbell K; Michael D; Black P. Using statistics to determine data adequacy for environmental policy decisions. LA-UR-98-3420. Los Alamos National Laboratory. 1998.

Keynes JM. *A Treatise on Probability*. Macmillan: London, 1921.

Knight K. On the bootstrap of the sample mean in the infinite variance case. *Annal Statist.* 1989; 17:1168–1173.

Krafft M; Kullgren A; Ydenius A; Tingvall C. Influence of crash pulse characteristics on whiplash associated disorders in rear impacts-crash recording in real life crashes. *Traffic Injury Prevention* 2002; 3:141–149.

Kumar S; Ferrari R; Narayan Y. Kinematic and electromyographic response to whiplash-type impacts. Effects of head rotation and trunk flexion: Summary of research. *Clinical Biomechanics* 2005; 20:553–568.

Lachin JM. Sample size determination. In *Encyclopedia of Biostatistics*, 5. Armitage P; Colton T. eds. John Wiley and Sons: Chichester, 1998; 3892–3903.

Lambert D. Robust two-sample permutation tests. *Ann. Statist.* 1985; 13:606–625.

Lang TA; Secic M. *How to Report Statistics in Medicine*. American College of Physicians: Philadelphia, 1997.

Lehmann EL. *Testing Statistical Hypotheses*. 2nd ed. New York: John Wiley and Sons, 1986. (pps 203–213 on robustness).

Lehmann EL. The Fisher Neyman-Pearson theories of testing hypotheses: One theory or two? *JASA* 1993; 88:1242–1249.

Lehmann EL; Casella G. *Theory of Point Estimation.* 2nd ed. Springer: New York, 1998.

Lehmann EL; D'Abrera HJM. *Nonparametrics: Statistical Methods Based on Ranks.* 2nd ed. McGraw-Hill: New York, 1988.

Leizorovicz A; Haugh MC; Chapuis F-R; Samama MM; Boissel J-P. Low molecular weight heparin in prevention of perioperative thrombosis. *BMJ* 1992; 305:913–920.

Lettenmaier DP. Space-time correlation and its effect on methods for detecting aquatic ecological change. *Canadian J. Fisheries Aquatic Science* 1985; 42:1391–1400. Correction 1986; 43:1680.

Lewis D; Burke CJ. Use and misuse of the chi-square test. *Psych Bull.* 1949; 46:433–489.

Liang KY; Zeger SL. Longitudinal data analysis using generalized linear models. *Biometrika* 1986; 73:13–22.

Lieberson S. *Making it Count.* University of California Press, Berkeley, 1985.

Light RJ; Pillemer DB. *Summing Up: The Science of Reviewing Research.* Harvard University Press: Cambridge; Massachusetts, 1984.

Lindley DV. The choice of sample size. *The Statistician* 1997; 46:129–138; 163–166.

Lindley D. The philosophy of statistics (with discussion). *The Statistician* 2000; 49:293–337.

Linnet K. Performance of Deming regression analysis in case of misspecified analytical error ratio in method comparison studies. *Clinical Chemistry* 1998; 44:1024–1031.

Linnet K. Necessary sample size for method comparison studies based on regression analysis. *Clinical Chemistry* 1999; 45:882–894.

Lissitz RW; Chardos S. A study of the effect of the violation of the assumption of independent sampling upon the type I error rate of the two group t-test. *Educat. Psychol. Measurement* 1975; 35:353–359.

Litière S. Alonso A; Mohlenberghs G. The impact of a misspecified random-effects distribution on the estimation and the performance of inferential procedures in generalized linear mixed models. *Statistics in Medicine* 2008; 27:3125–3144.

Little RJA; Rubin DB. *Statistical Analysis with Missing Data.* John Wiley and Sons: New York, 1987.

Loader C. *Local Regression and Likelihood.* Springer: NY, 1999.

Locke J. *Essay Concerning Human Understanding.* 4th ed. Prometheus Books, 1700.

Lonergan JF. *Insight: A Study of Human Understanding.* Univ. of Toronto Press, 1992.

Loo D. No paper trail left behind: The theft of the 2004 presidential election. http://www.projectcensored.org/newsflash/voter_fraud.html. 2005.

Lord FM. Statistical adjustment when comparing preexisting groups. *Psych Bull.* 1969; 72: 336–337.

Lovell DJ; Giannini EH; Reiff A; Cawkwell GD; Silverman ED; Nocton JJ; Stein LD; Gedalia A; Ilowite NT; Wallace CA; Whitmore J; Finck BK: The Pediatric Reumatology Collaborative Study Group. Etanercept in children with polyarticular juvenile rheumatoid arthritis. *New Engl. J. Med.* 2000; 342:763–769.

MacArthur RD; Jackson GG. An evaluation of the use of statistical methodology in the *Journal of Infectious Diseases. J. Infectious Diseases* 1984; 149:349–354.

Malone KM; Corbitt EM; Li S; Mann JJ. Prolactin response to fenuramine and suicide attempt lethality in major depression. *British J. Psychiatry* 1996; 168:324–329.

Mangel M; Samaniego FJ. Abraham Wald's work on aircraft survivability. *JASA* 1984; 79: 259–267.

Manly BFJ. *Randomization, Bootstrap and Monte Carlo Methods in Biology*. 2nd ed. London: Chapman and Hall, 1997.

Manly B; Francis C. Analysis of variance by randomization when variances are unequal. *Aust. New Zeal. J. Statist.* 1999; 41:411–430.

Maritz JS. *Distribution Free Statistical Methods*. 2nd ed. London: Chapman and Hall, 1996.

Martin RF. General Deming regression for estimating systematic bias and its confidence interval in method-comparison studies. *Clinical Chemistry* 2000; 46:100–104.

Matthews JNS; Altman DG. Interaction 2: Compare effect sizes not P values. *BMJ* 1996; 313: 808.

Mayo DG. *Error and the Growth of Experimental Knowledge*. University of Chicago Press, 1996.

McBride GB; Loftis JC; Adkins NC. What do significance tests really tell us about the environment? *Environ. Manage.* 1993; 17:423–432. (erratum. 19, 317).

McCullagh P; Nelder JA. *Generalized Linear Models*. 2nd ed. Chapman and Hall: London, UK, 1989.

McGuigan SM. The use of statistics in the *British Journal of Psychiatry*. British J. Psychiatry 1995; 167:683–688.

McKinney PW; Young MJ; Hartz A; Bi-Fong Lee M. The inexact use of Fisher's exact test in six major medical journals. *JAMA* 1989; 261:3430–3433.

Mehta CR; Patel NR. A hybrid algorithm for Fisher's exact test in unordered rxc contingency tables. *Commun. Statist.* 1986; 15:387–403.

Mehta CR; Patel NR; Gray R. On computing an exact confidence interval for the common odds ratio in several 2×2 contingency tables. *JASA* 1985; 80:969–973.

Mena EA; Kossovsky N; Chu C; Hu C. Inflammatory intermediates produced by tissues encasing silicone breast prostheses. *J. Invest. Surg.* 1995; 8:31–42.

Michaelsen J. Cross-validation in statistical climate forecast models. *J. Climate and Applied Meterorology* 1987; 26:1589–1600.

Mielke PW; Berry KJ. *Permutation Methods: A Distance Function Approach*. Springer: NY, 2001.

Mielke PW; Berry KJ. Permutation covariate analyses of residuals based on Euclidean distance. *Psychological Reports* 1997; 81:795–802.

Mielke PW; Berry KJ; Landsea CW; Gray WM. Artificial skill and validation in meteorological forecasting. *Weather and Forecasting* 1996; 11:153–169.

Mielke PW; Berry KJ; Landsea CW; Gray WM. A single sample estimate of shrinkage in meteorological forecasting. *Weather and Forecasting* 1997; 12:847–858.

Miller ME; Hui SL; Tierney WM. Validation techniques for logistic regression models. *Statist. Med.* 1991; 10:1213–1226.

Miller RG. Jackknifing variances. *Annals Math. Statist.* 1968; 39:567–582.

Miller RG. *Beyond Anova: Basics of Applied Statistics*. Wiley: NY, 1986.

Miyazaki Y; Terakado M; Ozaki K; Nozaki H. Robust regression for developing software estimation models. *J. Systems Software* 1994; 27:3–16.

Moher D; Cook DJ; Eastwood S; Olkin I; Rennie D; Stroup D. For the QUOROM Group. Improving the quality of reports of meta-analyses of randomised controlled trials: The QUOROM statement. *Lancet* 1999; 354:1896–1900.

Moiser CI. Symposium: The need and means of cross-validation, I: Problems and design of cross-validation. *Educat. Psych. Measure.* 1951; 11:5–11.

Montgomery DC; Myers RH. *Response Surface Methodology: Process and Product Optimization Using Designed Experiments.* Wiley: 1995.

Moore T. *Deadly Medicine: Why Tens of Thousands of Heart Patients Died in America's Worst Drug Disaster.* Simon; Schuster, 1995.

Morgan JN; Sonquist JA. Problems in the analysis of survey data and a proposal. *JASA* 1963; 58:415–434.

Morris RW. A statistical study of papers in the J. Bone and Joint Surgery, BR. *J. Bone and Joint Surgery BR* 1988; 70-B:242–246.

Morrison DE; Henkel RE. *The Significance Test Controversy.* Aldine: Chicago, 1970.

Mosteller F. Problems of omission in communications. *Clinical Pharmacology and Therapeutics* 1979; 25:761–764.

Mosteller F; Chalmers TC. Some progress and problems in meta-analysis of clinical trials. *Stat. Sci.* 1992; 7:227–236.

Mosteller F; Tukey JW. *Data Analysis and Regression: A second course in statistics.* Addison-Wesley, Menlo Park, 1977.

Moyé LA. *Statistical Reasoning in Medicine: The Intuitive P-Value Primer.* Springer: NY, 2000.

Mulrow CD. The medical review article: State of the science. *Ann. Intern. Med.* 1987; 106: 485–488.

Murray GD. Statistical guidelines for the *British Journal of Surgery.* *British J. Surgery* 1991; 78:782–784.

Murray GD. The task of a statistical referee. *British J. Surgery* 1988; 75:664–667.

Nelder JA; Wedderburn RWM. Generalized linear models. *JRSS A* 1972; 135:370–384.

Nester M. An applied statistician's creed. *Appl. Statist.* 1996; 45:401–410.

Neyman J. *Lectures and Conferences on Mathematical Statistics and Probability.* 2nd ed. Washington, Graduate School, U.S. Dept. of Agriculture, 1952.

Neyman J. Silver jubilee of my dispute with Fisher. *J. Operations Res. Soc. Japan* 1961; 3:145–154.

Neyman J. Frequentist probability and frequentist statistics. *Synthese* 1977; 36:97–131.

Neyman J; Pearson ES. On the testing of specific hypotheses in relation to probability a priori. *Proc. Cambridge Phil. Soc.* 1933; 29:492–510.

Neyman J; Pearson ES. On the problem of the most efficient tests of statistical hypotheses. *Phil. Trans. Roy. Soc. A* 1933; 231:289–337.

Neyman J; Scott EL. A theory of the spatial distribution of galaxies. *Astrophysical J.* 1952; 116: 144.

Nielsen-Gammon J. 2003. Sources of model error. http:p//www.met.tamu.edu/class/ ATMO151/tut/moderr/moderrmain.html.

Nunes T; Pretzlik U; Ilicak S. Validation of a parent outcome questionnaire from pediatric cochlear implantation. *J. Deaf Stud. Deaf Educ.* 2005; 10:330–356.

Nurminen M. Prognostic models for predicting delayed onset of renal allograft function. *Internet Journal of Epidemiology* 2003; 1:1.

Nurmohamed MT; Rosendaal FR; Bueller HR; Dekker E; Hommes DW; Vandenbroucke JP; et al. Low-molecular-weight heparin versus standard heparin in general and orthopaedic surgery: A meta-analysis. *Lancet* 1992; 340:152–156.

O'Brien PC. The appropriateness of analysis of variance and multiple-comparison procedures. *Biometrics* 1983; 39:787–788.

O'Brien P. Comparing two samples: Extension of the *t*, rank-sum, and log-rank tests. *JASA* 1988; 83:52–61.

Okano T; Kimura T; Tsugawa N; Oshio Y; Teraoka Y; Kobayashi T. Bioavailability of calcium from oyster shell electrolysate and DL-Calcium Lactate in Vitamin D-replete or Vitamin D-deficient rats. *J. Bone. Miner. Metab.* 1993; 11:S23–S32.

Oldham PD. A note on the analysis of repeated measurements of the same subjects. *J. Chron. Dis.* 1962; 15:969–977.

Olsen CH. Review of the use of statistics in *Infection and Immunity*. *Infection and Immunity* 2003; 71:6689–6692.

Osborne J; Waters E. Four assumptions of multiple regression that researchers should always test. *Practical Assessment, Research; Evaluation* 2002; 8(2).

Padaki PM. Inconsistencies in the use of statistics in horticultural research. *Hort. Sci.* 1989; 24:415.

Palmer RF; Graham JW; White EL; Hansen WB. Applying multilevel analytic strategies in adolescent substance use prevention research. *Prevent. Med.* 1998; 27:328–336.

Pankratz A. *Forecasting with Dynamic Regression Models*, New York: John Wiley and Sons, Inc. 1991.

Parkhurst DF. Arithmetic versus geometric means for environmental concentration data. *Environmental Science and Technology* 1998; 32:92A–98A.

Parkhurst DF. Statistical significance tests: Equivalence and reverse tests should reduce misinterpretation. *Bioscience* 2001; 51:1051–1057.

Perlich C; Provost F; Simonoff JS. Tree induction vs. logistic regression: A learning-curve analysis. *Journal of Machine Learning Research* 2003; 4:211–255.

Pesarin F. On a nonparametric combination method for dependent permutation tests with applications. *Psychotherapy and Psychosomatics* 1990; 54:172–179.

Pesarin F. *Multivariate Permutation Tests*. New York: Wiley, 2001.

Pettitt AN; Siskind V. Effect of within-sample dependence on the Mann–Whitney–Wilcoxon statistic. *Biometrika* 1981; 68:437–441.

Phipps MC. Small samples and the tilted bootstrap. *Theory of Stochastic Processes* 1997; 19: 355–362.

Picard RR; Berk KN. Data splitting. *American Statistician* 1990; 44:140–147.

Picard RR; Cook RD. Cross-validation of regression models. *JASA* 1984; 79:575–583.

Pierce CS. *Values in a University of Chance*. Wiener PF. (ed.), Doubleday Anchor Books: NY, 1958.

Pilz J. *Bayesian Estimation and Experimental Design in Linear Regression Models*. 2nd ed. Wiley: New York, 1991.

Pinelis IF. On minimax risk. *Theory Prob. Appl.* 1988; 33:104–109.

Pitman EJG. Significance tests which may be applied to samples from any population. *Roy. Statist. Soc. Suppl.* 1937; 4:119–130, 225–232.

Pitman EJG. Significance tests which may be applied to samples from any population. Part III. The analysis of variance test. *Biometrika* 1938; 29:322–335.

Poole C. Beyond the confidence interval. *Amer. J. Public Health* 1987; 77:195–199.

Poole C. Low p-values or narrow confidence intervals: Which are more durable? *Epidemiology* 2001; 12:291–294.

Praetz P. A note on the effect of autocorrelation on multiple regression statistics. *Australian J. Statist.* 1981; 23:309–313.

Proschan MA; Waclawiw MA. Practical guidelines for multiplicity adjustment in clinical trials. *Controlled Clinical Trials* 2000; 21:527–539.

Rabe-Hesketh S; Skrondal A. *Multilevel and Longitudinal Modeling Using Stata.* Stata Press, College Station, TX, 2008.

Ravnskov U. Cholesterol lowering trials in coronary heart disease: Frequency of citation and outcome. *BMJ* 1992; 305:15–19.

Rea LM; Parker RA; Shrader A. *Designing and Conducting Survey Research: A Comprehensive Guide.* Jossey-Bass. 2nd ed. 1997.

Redmayne M. Bayesianism and Proof, in *Science in Court*, M. Freeman, Reece, H. eds., Ashgate: Brookfield, MA, 1998.

Reichenbach H. *The Theory of Probability.* University of California Press: Berkeley, 1949.

Rencher AC; Pun F-C. Inflation of R^2 in best subset regression. *Technometrics* 1980; 22:49–53.

Rice SA; Griffin JR. The hornworm assay: Useful in mathematically-based biological investigations. *American Biology Teacher* 2004; 66:487–491.

Rosenbaum PR. *Observational Studies.* Springer, 2nd ed. 2002.

Rosenberger W; Lachin JM. *Randomization in Clinical Trials: Theory and Practice.* John Wiley and Sons, New York, 2002.

Rothman KJ. Epidemiologic methods in clinical trials. *Cancer* 1977; 39:1771–1775.

Rothman KJ. No adjustments are needed for multiple comparisons. *Epidemiology* 1990; 1:43–46.

Rothman KJ. Statistics in nonrandomized studies, *Epidemiology* 1990; 1:417–418.

Roy J. Step-down procedure in multivariate analysis. *Ann. Math. Stat.* 1958; 29-4:1177–1187.

Royall RM. *Statistical Evidence: A Likelihood Paradigm.* Chapman and Hall: New York, 1997.

Rozeboom W. The fallacy of the null hypothesis significance test. *Psychol. Bull.* 1960; 57:416–428.

Rozen TD; Oshinsky ML; Gebeline CA; Bradley KC; Young WB; Shechter AL; Silberstein SD. Open label trial of coenzyme Q10 as a migraine preventive. *Cephalalgia* 2008; 22:137–141.

Salmaso L. Synchronized permutation tests in 2^k factorial designs. *Int. J. Non Linear Model. Sci. Eng.* 2002; 3.

Savage LJ. *The Foundations of Statistics.* Dover Publications, 1972.

Saville DJ. Multiple comparison procedures: The practical solution. *American Statistician* 1990; 44, 174–180.

Saville DJ. Basic statistics and the inconsistency of multiple comparison procedures. *Canadian J. Exper. Psych.* 2003; 57(3):167–175.

Schmidt FL. Statistical significance testing and cumulative knowledge in psychology: Implications for training of researchers. *Psychol. Meth.* 1996; 1:115–129.

Schenker N. Qualms about bootstrap confidence intervals. *JASA* 1985; 80:360–361.

Schor S; Karten I. Statistical evaluation of medical manuscripts. *JASA* 1966; 195:1123–1128.

Schroeder YC. The procedural and ethical ramifications of pretesting survey questions. *Amer. J. of Trial Advocacy* 1987; 11:195–201.

Schulz KF. Randomised trials, human nature, and reporting guidelines. *Lancet* 1996; 348:596–598.

Schulz KF. Subverting randomization in controlled trials. *JAMA* 1995; 274:1456–1458.

Schulz KF; Chalmers I; Hayes R; Altman DG. Empirical evidence of bias. Dimensions of methodological quality associated with estimates of treatment effects in controlled trials. *JAMA* 1995; 273:408–412.

Schulz KF, Grimes DA. Blinding in randomized trials: Hiding who got what. *Lancet* 2002; 359:696–700.

Seidenfeld T. *Philosophical Problems of Statistical Inference*. Reidel: Boston, 1979.

Selike T; Bayarri MJ; Berger J. Calibration of *p*-values for testing precise null hypotheses. *Amer. Statist.* 2001; 55:62–71.

Selvin H. A critique of tests of significance in survey research. *Amer. Soc. Rev.* 1957; 22: 519–527.

Senn S. A personal view of some controversies in allocating treatment to patients in clinical trials. *Statist. Med.* 1995; 14:2661–2674.

Shao J; Tu D. *The Jacknife and the Bootstrap*. New York: Springer, 1995.

Sharp SJ; Thompson SG; Altman DG. The relation between treatment benefit and underlying risk in meta-analysis. *BMJ* 1996; 313:735–738.

Sharp SJ; Thompson SG. Analysing the relationship between treatment effect and underlying risk in meta-analysis: Comparison and development of approaches. *Statist. Med.* 2000; 19:3251–3274.

Shuster JJ. *Practical Handbook of Sample Size Guidelines for Clinical Trials*. CRC: Boca Raton, 1993.

Simes RJ. Publication bias: The case for an international registry of clinical trials. *J. Clinical Oncology* 1986; 4:1529–1541.

Simpson JM; Klar N; Donner A. Accounting for cluster randomization: A review of primary prevention trials; 1990 through 1993. *Am. J. Public Health* 1995; 85:1378–1383.

Skrondal A; Rabe-Hesketh S. *Generalized Latent Variable Modeling: Multilevel, Longitudinal and Structural Equation Models*. Chapman and Hall/CRC. Boca Raton, FL, 2004.

Smeeth L; Haines A; Ebrahim S. Numbers needed to treat derived from meta-analysis— sometimes informative; usually misleading. *BMJ* 1999; 318:1548–1551.

Smith GD; Egger M. Commentary: Incommunicable knowledge? Interpreting and applying the results of clinical trials and meta-analyses. *J. Clin. Epidemiol.* 1998; 51:289–295.

Smith GD; Egger M; Phillips AN. Meta-analysis: Beyond the grand mean? *BMJ* 1997; 315:1610–1614.

Smith TC; Spiegelhalter DJ; Parmar MKB. Bayesian meta-analysis of randomized trials using graphical models and BUGS. In *Bayesian Biostatistics*. Berry DA; Stangl DK. eds. Marcel Dekker: New York, 1996; 411–427.

Snee RD. Validation of regression models: Methods and examples. *Technometrics* 1977; 19: 415–428.

Sox HC; Blatt MA; Higgins MC; Marton KI. *Medical Decision Making*. Butterworth and Heinemann: Boston, 1988.

Spiegelhalter DJ. Probabilistic prediction in patient management. *Statist. Med.* 1986; 5: 421–433.

Sterne JAC; Smith GD; Cox DR. Sifting the evidence—what's wrong with significance tests? Another comment on the role of statistical methods. *BMJ* 2001; 322:226–231.

Stewart L; Parmar M. Meta-analysis of the literature or of individual patient data: Is there a difference? *Lancet* 1993; 341:418–422.

Stöckl D; Dewitte K; Thienpont LM. Validity of linear regression in method comparison studies: Is it limited by the statistical model or the quality of the analytical input data? *Clinical Chemistry* 1998; 44:2340–2346.

Stockton CW; Meko DM. Drought recurrence in the Great Plains as reconstructed from long-term tree-ring records. *J. of Climate and Applied Climatology* 1983; 22:17–29.

Stone M. Cross-validatory choice and assessment of statistical predictions. *JRSS B* 1974; 36: 111–147.

Su Z; Adkison MD; Van Alen BW. A hierarchical Bayesian model for estimating historical salmon escapement and escapement timing. *Canadian J. Fisheries and Aquatic Sciences* 2001; 58:1648–1662.

Subrahmanyam M. A property of simple least squares estimates. *Sankha* 1972; 34B:355–356.

Sukhatme BV. A two sample distribution free test for comparing variances. *Biometrika* 1958; 45:544–548.

Suter GWI. Abuse of hypothesis testing statistics in ecological risk assessment. *Human and Ecological Risk Assessment* 1996; 2:331–347.

Teagarden JR. Meta-analysis: Whither narrative review? *Pharmacotherapy* 1989; 9:274–284.

Tencer AF; Sohail M; Kevin B. The response of human volunteers to rear-end impacts: The effect of head restraint properties. *Spine* 2001; 26:2432–2440.

Therneau TM; Grambsch PM. *Modeling Survival Data*. Springer: New York. 2000.

Thompson SG. Why sources of heterogeneity in meta-analysis should be investigated. *BMJ* 1994; 309:1351–1355.

Thompson SK; Seber GAF. *Adaptive Sampling*. Wiley, 1996.

Thorn MD; Pulliam CC; Symons MJ; Eckel FM. Statistical and research quality of the medical and pharmacy literature. *American J. Hospital Pharmacy* 1985; 42:1077–1082.

Tiku ML; Tan WY; Balakrishnan N. *Robust Inference*. New York and Basel: Marcel Dekker, 1990.

Tokita A; Maruyama T; Mori T; Hayashi M; Nittono H; Yabuta K. (1993) Intestinal absorption of AACa in bile duct ligated rats. *J. Bone Miner. Met.* 1993; 11(S2):S53–S55.

Torri V; Simon R; Russek-Cohen E; Midthune D; Friedman M. Statistical model to determine the relationship of response and survival in patients with advanced ovarian cancer treated with Chemotherapy. *J. Nat. Cancer Institut.* 1992; 84:407–414.

Tsai C-C; Chen Z-S; Duh C-T; Horng FW. Prediction of soil depth using a soil-landscape regression model: A case study on forest soils in southern Taiwan. *Proc. Natl. Sci. Counc. ROC(B)*. 2001; 25:34–39.

Tu D; Zhang Z. Jackknife approximations for some nonparametric confidence intervals of functional parameters based on normalizing transformations. *Comput. Statist.* 1992; 7:3–5.

Tufte ER. *The Visual Display of Quantitative Information*. Graphics Press: Cheshire Ct. 1983.

Tufte ER. *Envisioning Data. Graphics Press*. Graphics Press: Cheshire Ct. 1990.

Tukey JW. *Exploratory Data Analysis*. Addison-Wesley: Reading MA, 1977.

Tukey JW. The philosophy of multiple comparisons. *Statist. Sci.* 1991; 6:100–116.

Tukey JW; McLaughlin DH. Less vulnerable confidence and significance procedures for location based on a single sample; Trimming/Winsorization 1. *Sankhya* 1963; 25:331–352.

Turner RB; Bauer R; Woelkart K; Hulsey TC; Gangemi JD. An evaluation of echinacea angustifolia in experimental rhinovirus infections. *New England Journal Medicine* 2005; 353: 341–348.

Tversky A; Kahneman D. Belief in the law of small numbers. *Psychol. Bull.* 1971; 76:105–110.

Toutenburg H. *Statistical Analysis of Designed Experiments*. 2nd ed. Springer-Verlag: New York, 2002.

Tyson JE; Furzan JA; Reisch JS; Mize SG. An evaluation of the quality of therapeutic studies in perinatal medicine. *J. Pediatrics* 1983; 102:10–13.

United States Environmental Protection Agency. *Data Quality Assessment: Statistical Methods for Practitioners* EPA QA/G-9S EPA. D.C, 2006.

Vaisrub N. Manuscript review from a statisticians perspective. *JAMA* 1985; 253:3145–3147.

van Belle G. *Statistical Rules of Thumb*. Wiley: New York, 2002.

Venn J. *The Logic of Chance*. MacMillan: London, 1888.

Victor N. The challenge of meta-analysis: Discussion. *J. Clin. Epidemiol.* 1995; 48:5–8.

Wainer H. Rounding tables. *Chance* 1998; 11:46–50.

Wainer H. *Visual Revelations: Graphical Tales of Fate and Deception From Napoleon Bonaparte To Ross Perot*. Springer: NJ, 1997.

Wainer H. *Graphic Discovery: A Trout in the Milk and Other Visual Adventures*. Princeton University Press, 2004.

Watterson IG. Nondimensional measures of climate model performance. *Int. J. Climatology* 1966; 16:379–391.

Weeks JR; Collins RJ. Screening for drug reinforcement using intravenous self-administration in the rat. In: Bozarth MA. (ed.), *Methods of Assessing the Reinforcing Properties of Abused Drugs*. New York: Springer-Verlag, 1987; 35–43.

Weerahandi S. *Exact Statistical Methods for Data Analysis*. Springer Verlag: Berlin, 1995.

Weisberg S. *Applied Linear Regression*. 2nd ed; John Wiley; New York, 1985.

Welch BL. On the z-test in randomized blocks and Latin squares. *Biometrika* 1937; 29:21–52.

Welch GE; Gabbe SG. Review of statistics usage in the *American J. Obstetrics and Gynecology. American J. Obstetrics and Gynecology* 1996; 175:1138–1141.

Westfall DH; Young SS. *Resampling-Based Multiple Testing: Examples and Methods for p-value Adjustment*. New York: John Wiley, 1993.

Westgard JO. Points of care in using statistics in method comparison studies. *Clinical Chemistry* 1998; 44:2240–2242.

Westgard JO; Hunt MR. Use and interpretation of common statistical tests in method comparison studies. *Clin. Chem.* 1973; 19:49–57.

White SJ. Statistical errors in papers in the British J. Psychiatry. *British J. Psychiatry* 1979; 135:336–342.

Whitehead J. Sample size calculations for ordered categorical data. *Statistics in Medicine* 1993; 12:2257–2271. 1994; 13:871.

Wilkinson L. *The Grammar of Graphics.* Springer-Verlag: New York, 1999.

Wilks DS. *Statistical Methods In The Atmospheric Sciences.* Academic Press, 1995.

Willick JA. Measurement of galaxy distances. In *Formation of Structure in the Universe,* Dekel A; Ostriker J. eds. Cambridge University Press, 1999.

Wilson JW; Jones CP; Lundstrum LL. Stochastic properties of time-averaged financial data: Explanation and empirical demonstration using monthly stock prices. *Financial Review* 2001; 36:3.

Wise TA. Understanding the farm problem: Six common errors in presenting farm statistics. http://www.ase.tufts.edu/gdae/Pubs/wp/05-02TWiseFarmStatistics.pdf. 2005.

Wu CFJ. Jackknife, bootstrap, and other resampling methods in regression analysis (with discuss). *Annals Statist.* 1986; 14:1261–1350.

Wu DM. Alternative tests of independence between stochastic regressors and disturbances. *Econometrica* 1973; 41:733–750.

Wulf HR; Andersen B; Brandenhof P; Guttler F. What do doctors know about statistics? *Statistics in Medicine* 1987; 6:3–10.

Yandell BS. *Practical Data Analysis for Designed Experiments.* Chapman and Hall: London, 1997.

Yoccoz NG. Use, overuse, and misuse of significance tests in evolutionary biology and Ecology. *Bull. Ecol. Soc. Amer.* 1991; 72:106–111.

Yoo S-H. A robust estimation of hedonic price models: Least absolute deviations estimation. *Applied Economics Letters* 2001; 8:55–58.

Young A. Conditional data-based simulations: Some examples from geometric statistics. *Int. Statist. Rev.* 1986; 54:1–13.

Zhou X-H; Gao S. Confidence intervals for the log-normal mean. *Statist. Med.* 1997; 17: 2251–2264.

AUTHOR INDEX

Adams DC, 114
Adkins NC, 135
Adkison MD, 114
Aickin M, 97, 132, 147
Albers W, 77
Alonso A, 222
Altman DG, 11, 15, 66, 73, 97, 99, 132, 135
Aly E-E AA, 86
Andersen B, 47
Anderson DR, 16
Anderson PJ, 111
Anderson S, 79, 136
Anderson SL, 85, 99
Anscombe F, 16
Archfield SA, 209
Avram MJ, 97

Bacchetti P, 99
Badrick TC, 97
Bailar JC, 136
Bailey KR, 111
Bailor AJ, 85
Baker RD, 86, 88, 95–96

Balakrishnan N, 85
Barankin E, 108
Barbui C, 44
Barnard GA, 225
Barnston AG, 231
Barrodale I, 195
Batanero C, 142, 146, 209
Bauer R, 119
Bayarri MJ, 25, 109
Bayes T, 103
Begg C, 111
Begg CB, 136
Bent GC, 209
Berger VC, 21, 45, 53, 97, 109, 111, 114, 118
Berger JO, 47, 57, 66, 100
Berk KN, 228
Berkeley G, 25, 100
Berkey C, 111
Berkson J, 25
Berlin J, 111
Berlin JA, 114
Berry DA, 25, 107, 113, 186, 203
Berry KJ, 114

Best R, 207
Bickel PJ, 66, 77
Bishop G, 28, 29
Bithell J, 63
Black P, 52
Bland JM, 66, 97
Blatt MA, 25
Block G, 28
Bly RW, 30, 47
Blyth CR, 25
Bond J, 11
Boomsma A, 231
Bothun G, 7, 187
Box GEP, 85, 96, 99, 217
Bradley JV, 113
Brand C, 208
Breiman L, 213
Brockwell PJ, 227
Brown MB, 87
Browne MW, 227
Buchanan-Wollaston H, 25
Burn DA, 173
Burnham KP, 16
Buyse M, 76

Cade B, 186, 200
Callaham ML, 134
Campbell K, 52
Campbell MJ, 99
Camstra A, 231
Canty AJ, 103, 113
Capaldi DM, 121
Cappuccio FP, 28
Carlin BP, 114
Carlson J, 207
Carpenter J, 63
Carroll RJ, 27, 66, 202
Casella G, 47, 57, 66
Chalmers TC, 44, 111, 114
Chernick MR, 34, 102
Cherry S, 97
Chiles JR, 15
Choi BCK, 31
Christophi C, 118
Chu KC, 22
Clemen RT, 25, 113
Cleveland WS, 167, 172
Cochran WG, 47

Cody R, 223
Cohen J, 25
Collins RJ, 78–79
Conan-Doyle A, 21
Conover WJ, 81, 85
Converse JM, 30, 47
Cooper MM, 11
Cook RD, 226
Copas JB, 25, 47
Cornfield J, 94
Cox DR, 25, 97, 100, 136
Cummings P, 7
Cupples LA, 97

D'Abrera HJM, 95
Dar R, 11, 97
Davis RA, 227
Davision AC, 103, 191
Day S, 118
Dean CW, 132, 137
DeGroot MH, 25
Delucchi KL, 97
DeMets DL, 192
Dewitte K, 198, 203
Diaconis P, 8
Diciccio TJ, 113
Dixon PM, 79, 99
Djulbegovic B, 44
Duggan TJ, 132, 137
Durbin J, 219
Dyke G, 3, 67, 97, 125

Easterbrook PJ, 111
Ebrahim S, 114
Eckel FM, 11
Ederer F, 44
Edwards W, 109
Efron B, 63, 66, 86, 113
Egger M, 15, 110–114
Ehrenberg ASC, 125–126
Ellis SP, 196, 203
Elwood JM, 47, 97
Exner DV, 45
Eysenbach G, 114

Fears TR, 22
Feinstein AR, 132, 135
Feller W, 128

Felson DT, 99, 111
Feng Z, 82–84
Fergusson D, 119
Ferrari R, 28
Fienberg SE, 99
Fink A, 47
Finney DJ, 99
Firth D, 218
Fisher N, 113
Fisher RA, 47, 73
Flatman RJ, 97
Fleischer AB,
Fleming TR, 183
Fligner MA, 85
Forsythe AB, 87
Fowler FJ, 30, 47
Francis C, 80
Frank D, 21
Freedman DA, 191–192, 205
Friedman JH, 213
Friedman LM, 192
Friedman M, 99
Friendly M, 157
Fujita T, 42–43, 52, 121, 140
Fukada S, 46
Furberg CD, 192
Furzan JA, 11

Gabbe SG, 99
Gail MH, 84
Gallant AR, 202
Gao S, 189
Gardner MJ, 99, 132
Garthwaite PH, 101
Gastwirth JL, 99
Geary RC, 89
Gensler H, 97, 132, 142
George SL, 11, 99
Gigerenzer G, 96
Gillett R, 114
Gine E, 113
Glantz S, 11
Glass GV, 99
Glass KC, 119
Godino JD, 142, 146, 209
Goldberger AS, 205
Gong G, 206, 215
Gonzales C, 141

Gonzales GF, 141
Good IJ, 108, 114
Good PI, 47, 75, 81, 86–90, 95–96,
    100–104, 189, 203, 211
Goodman SN, 11, 99, 108
Gore S, 99
Götzsche PC, 111
Graham JW, 114
Grambsch PM, 216
Grant A, 135
Graumlich L, 226
Gray R, 73
Green PJ, 191
Greene HL, 38
Greenland S, 108, 135
Griffin JR, 147
Grimes DA, 47
Gurevitch J, 114
Guthery FS, 25
Gutiérrez-Jaimez RG, 142,146

Hagood MJ, 14
Haines A, 114
Hall P, 63, 82, 113
Hallock KF, 200
Hansen WB, 114
Hardin J, 173, 191, 216, 222
Harley SJ, 114
Harrell FE, 216
Hartel G, 94
Hauck WW, 79, 136
Hausman JA, 219
Hedges LV, 113, 114
Henkel RE, 25
Hershey D, 147
Hertwig R, 16
Heywood J, 214
Higgins MC, 25
Hilbe JM, 173, 191, 216, 222
Hilton J, 81
Hinkley DV, 103
Hodges JS, 109
Horwitz RI, 15, 11, 113, 136
Hosmer DW, 216
Hout M, 207
Hsu JC, 97, 132, 142
Huber PJ, 66
Hubley AM, 135

Hui Y, 168, 231
Hume D, 25, 100
Hunt MR, 99
Hunter JE, 25, 99, 137
Hurlbert SH, 47
Hutchon DJR, 114
Husted JA, 55

Ilicak S, 52
International Committee of Medical
    Journal Editors, 99
International Study of Infarct
    Survival Collaborative Group, 8
Ivanova A, 21

Jackson GG, 11, 99
Jagers P, 93
Jaimez RG, 209
Jefferson BL, 159
Jennison C, 47
Johnson DH, 135
Johnson ME, 85
Johnson MM, 85
Jones CP, 200
Jones IG, 99
Jones LV, 131, 135
Jones SK, 113

Kadane IB, 108
Kanarek MS, 184, 188
Kaplan J, 25
Karten I, 99
Kass R, 113
Kaye DH, 107
Kelly E, 52
Kevin B, 28
Keynes JM, 103
Killeen TJ, 85
Knebel F, 139
Knight K, 113
Koenker R, 200
Koepsell TD, 7
Kosecoff JB, 47
Krafft M, 28
Kumar S, 28

Lachin JM, 47
Lambert D, 77, 114
Lang TA, 121, 128, 156

Lee KL, 216
Lehmann EL, 25, 47, 59, 66, 72,
    77–85, 90, 95
Leizorovicz A, 111
Lemeshow SL, 216
Li HG, 25, 47
Liang KY, 219, 221
Lieberson S, 99
Light RJ, 99
Lindley DV, 25, 47
Linnet K, 47, 197
Litière S, 222
Liu CY, 34
Loader C, 191
Locke J, 25, 100
Loftis JC, 135
Lonergan JF, 25, 99–100
Loo D, 207
Louis TA, 114
Lovell DJ, 45
Lundstrum LL, 200
Lunneborg CE, 89
Lusk JJ, 25

Ma CW, 85
MacArthur RD, 99
Machin D, 99
Malone KM, 196
Mangel M, 39, 98
Mangels L, 207
Manly BFJ, 80, 100, 114
Maritz JS, 66, 77, 100, 114
Mark SD, 216
Martin RF, 203
Marton KI, 25
Matthews JNS, 132
Mayo DG, 25
McBride GB, 135
McCullagh P, 223
McGill ME, 167
McGuigan SM, 99
McHale D, 220
McKinney PW, 20, 99
McLaughlin DH, 96
Meenan RF, 97
Mehta CR, 73
Meko DM, 226
Mena EA, 180
Michael D, 52

Michaelsen J, 231
Mieike PW, 114, 186, 203, 230–231
Miller RG, 85, 99, 231
Miyazaki Y, 203
Mize SG, 11
Moher D, 114
Mohlenberghs G, 222
Moiser CI, 228
Montgomery DC, 47
Moore T, 38
Morgan JN, 216
Morris RW, 11, 66
Morrison DE, 25
Mosteller F, 81, 111, 114, 136, 187,
     192, 216
Moye L, 38, 79, 100
Mulrow CD, 99
Murray GD, 99
Myers RA, 47, 114

Narayan Y, 28
Nelder JA, 216, 223
Nester M, 135
Neyman J, 17–19, 25, 47, 99, 123, 136
Nielsen-Gammon J, 208
Nunes T, 52
Nurminen M, 216
Nurmohamed MT, 111

O'Brien PC, 80, 142
Oikin I, 114
Okano T, 140
Olsen CH, 67, 142
Olshen RA, 213
Omer H, 11, 97
Osborne J, 216

Padaki PM, 99
Palmer RF, 114
Pankratz A, 186
Parker RA, 47
Parkhurst DF, 135–136
Parmar M, 114
Parzen E, 149
Patel NR, 73
Patterson GR, 121
Pearson ES, 17–19, 25
Peckham PD, 99
Perlich C, 216

Permutt T, 21
Pesarin F, 80–81, 100, 114
Pettitt AN, 99
Phillips AN, 110, 114
Phipps MC, 103
Picard RR, 226, 228
Piedbois P, 76
Pillemer DB, 99
Pilz J, 60
Pinelis IF, 60
Pitman EJG, 112, 186
Plato, 98
Poole C, 132, 136
Presser S, 30, 47
Pretzlik U, 52
Provost F, 216
Pulliam CC, 11
Pun F-C, 230

Rabe-Hesketh S, 223
Raftery A, 113
Rea LM, 47
Redmayne M, 103
Reichenbach H, 25, 100
Reisch JS, 11
Rencher AC, 216, 230
Rice SA, 147
Richards L, 186
Roberts EM, 119
Roberts FDK, 195
Romano JP, 100, 113
Rosenbaum PR, 47
Rosenberg MS, 114
Rosenberger W, 47
Rosenthal R, 11
Rothman KJ, 100, 135, 142
Roy J, 205
Rozen tD, 119
Rozeboom W, 25
Rubin H, 99
Ruppert D, 66
Rytter EC, 99

Sa ER, 114
Salmaso L, 95, 100
Salsburg D, 81
Samaniego FJ, 39, 98
Sanders JR, 99
Savage LJ, 25

Saville DJ, 142
Schenker N, 113
Schmidt FL, 25, 99, 137
Schor S, 99
Schroeder YC, 30, 47
Schultz KF, 45, 47
Scott EL, 123
Seber GAP, 47
Secic M, 121, 128, 136
Selike T, 25, 109
Selvin H, 135
Seidenfeld T, 25
Senn S, 118
Serlin OH, 97
Shao J, 66, 229–231
Sharp SJ, 114
Shi S, 103
Shrader A, 47
Shuster JJ, 47
Silverman BW, 191
Simes RJ, 114
Simonoff JS, 216
Siskind V, 99
Skrondal A, 223
Smeeth L, 114
Smith GD, 15, 25, 100, 110, 113–114
Snell EJ, 191
Sohail M, 28
Sonquist JA, 216
Sox HC, 25
Spiegelhalter DJ, 114
Stangi DK, 113
Stefanski LA, 66
Sterne JAC, 25, 100
Stewart L, 114
Still AW, 95
Stöckl D, 198, 203
Stockton CW, 226
Stone CJ, 213
Stone M, 225
Su Z, 114
Subrahmanyam M, 228
Sukhatme BV, 85
Suter GWI, 135
Symons MJ, 11

Tabor B, 218
Talbot M, 28, 29

Tarone RE, 22
Tatem AJ, 188
Teagarden JR, 113
Tencer AF, 28
Therneau TM, 198, 216
Thienpont LM, 203
Thompson SG, 111, 114
Thompson SK, 47
Thompson WL, 16
Thorn MD, 11
Tiao GC, 96
Tibshirani R, 63, 66
Tierney, 231
Todd PM, 16
Tokita A, 126, 166
Toutenburg H, 47
Tribe L, 104
Trzos R, 21
Tsai C-C, 226
Tu D, 64, 66, 229, 231
Tufte ER, 136, 166, 172
Tukey JW, 81, 84, 94, 96, 99, 135, 142,
    172, 187, 192, 216, 220
Turnbull BW, 47
Turner RB, 119
Tyson JE, 11

Vaisrub N, 99
van Alen BW, 114
van Belle G, 125–126
van den Dool HM, 231
van Zwet WR, 77
Venn J, 25, 100
Vickers A, 44
Victor N, 111
von Neumann, 202

Wainer H, 173
Wald A, 37
Waters E, 216
Watterson IG, 231
Wedderburn RWM, 216
Weeks JR, 78–79
Weerahandi S, 80
Weisberg S, 231
Welch GE, 85, 99

Westfall DH, 97, 132
Westgard JO, 99, 193, 199
White EL, 114
White SJ, 95, 99
Whitehead J, 33
Wilkinson L, 166, 172
Wilks DS, 230
Willick JA, 134
Wilson JW, 200
Wilson SR, 63, 82
Wise TA, 140
Wu CFJ, 113, 229
Wu DM, 219

Xie F, 95

Yoccuz NG, 99, 132, 135
Yoo S-H, 59
Young A, 82, 97, 132

Zeger SL, 219, 221
Zhang Z, 64
Zhou X-H, 189
Zinn J, 113
Zumbo BD, 200

SUBJECT INDEX

a priori distribution, 105–7
a priori probability, 61, 104, 108
Acceptance region, 18, 132
Accuracy vs. precision, 127, 233
Additive vs. multiplicative, 46
Agronomy, 21, 38, 41, 183
AIDs, 183
Algorithms, 193, 202, 217, 230
Allocation (of treatment),
    see Treatment allocation
Aly's statistic, 86
Analysis of variance, 21, 74, 89, 92–94,133
Angiograms, 31, 120
Animals, 7, 38
Antibodies, 28
ARIMA, 227
Arithmetic vs. geometric mean, 124–125
Aspirin, 8, 44
Association
  spurious, 8
  versus causation, 131
Assumptions, 4, 69, 96, 192, 219
Astronomy, 40, 123, 134
Asymptotic,
  approximation, 72, 83, 85, 230

properties, 88, 102, 222
relative efficiency (ARE), 59
Audit, 187
Authors, 139
Autocorrelation, 135, 215
Autoregressive process, see time series
Axis
  label, 155
  range, 154–155

Bacteria, 125, 183
Balanced design,
Bar chart, 122, 151–152, 164–165
Baseline, 42, 52, 76
Bayes
  analysis, 111
  factor, 108–110
  Theorem, 103–104
Behrens–Fisher problem, 102
Bias
  estimation, 102, 197
  measurement, 210
  recognizing, 133
  sample, 7,
  selection, 53, 55, 110, 135

Bias (*Continued*)
  sources, 133
  systematic error, 135
  time, 76
  under-reporting, 98, 134
Binomial trials, 34, 71, 122
Blinding, 44, 52, 119, 141
Blocks, 41, 74
Blood, 30
Blood pressure, 9, 15, 28, 183
Bonferroni correction, 97
Bootstrap, 102
  BCA, 64
  Estimates, 35
  limitations, 162–163
  nonparametric, 61
  parametric, 65
  percentile/primitive, 61
  sample, 130
  test, 80–81
  tilted, 103
  validation, 228
Box and whiskers plot, 53, 124, 129
Boyle's Law, 5

Cancer, 21, 23, 69, 107, 184
Caption, 166
CART, 211–214
Case controls, 47
Categorical data, 72
  2 × 2 table, 73, 122
  stratified tables, 73
  ordered categories, see Ordinal data
Cause and effect, 192, 214
Censored data, 74–75
Census, 187
Central limit theorem, 128–130
Chisquare
  statistic, 73
  statistic vs. distribution, 72, 75
Circadian rhythm, 91
Classification and regression tree,
    see CART
Climatology, 226
Clinical
  chemistry, 197
  significance, 79
  trials, 9, 24, 38, 40, 45, 107, 130
Cluster, 81

Cold fusion, 8
Coefficients
  estimating, 185
Computer
  output, 117, 133, 146
  simulation, see Simulations
Confidence interval, 102, 131, 147
Confounded effects, 92, 96, 135, 207
Contagion, 218
Contingency table, 19, 72–73, 142
Contour plot, 160
Contrasts, 93, 126
Controls, 41, 43, 52
  positive, 43
Correlation,
  Pearson, 91
  spurious, 183, 189, 225
  structure, 221–222
  vs independence, 91
  vs. slope, 198
Corticosteroids, 7
Cost-benefit analysis, 23
Counts, 218
Covariances, 82
Covariates, 75–76, 84
Criminology, 182
Crossover designs, 45, 95, 112
Curvefitting, 190
Cutoff value, 64

Data
  aggregation, 140
  baseline, 140
  categorical, 72, 122
  censored, 73
  collection, 27
  correlated, 217ff
  found, 16
  ordinal, 125
  panel, 218
  quality, 51
  subjective, 91
  time-to-event, 73
  types, 233
Data mining, 212
Deaths, 10, 28, 38, 111, 121, 145, 184
Decimal places, 125, 147
Decision tree, see CART
Decision theory, 23

Deduction, 21
Degrees of freedom, 222
Delete-one, 187
Dependent observations, 81
Descriptive statistics, 122, 141
Deterministic vs. stochastic, 233
Dimensions, 149
Discrimination, 69, 183
Disease process, 17
Dispersion, 86, 127, 141, 200
Distribution
  a priori, 109
  beta, 108
  bivariate, 78
  cumulative, 234
  empirical, 234
  exponential, 127
  F, 85, 89, 112
  function, 33
  mixture, 127
  multiparameter, 70
  multivariate normal, 78
  nonsymmetric, 59
  normal, 64, 128
  Poisson, 9, 72, 79, 127
  sampling, 130
  symmetric, 59, 103, 130
  uniform, 108, 128
Diurnal rhythm, see Circadian
Domain expert, 33, 134
Dose
  maximum tolerable, 15
  response, 110
Dropouts, 7, 32, 120
Drugs, 123
Dynamic models, 208

Ecological fallacy, 201
Effect size, 117
Elections, 201
Eligibility, 120
Empirical distribution, 234
Emissions, 24
Endpoints, 28, 38
Epidemiology, 110
Equivalence, 79
Error
  common, 3
  measurement, 30, 57, 191

prediction, 230
probabilities, 16
terms, 112, 186
Estimate, 58
  admissible, 65
  consistent, 58, 219
  efficient, 58, 219
  interval vs. point, 61
  minimax, 60
  minimum loss, 60
  minimum variance, 60, 185
  panel, 221
  plugin, 60
  population-averaged, 220
  ratio, 130
  robust, 59
  unbiased, 60, 185
  variance, 220
Experimental design, 41
  unbalanced, 46, 83, 90, 94
Experimental unit, 40
Exploratory analysis, 125
Extrapolate, 178, 226, 228

Factor analysis, 142, 208
Factorial experiments, 95
False dimension, 151–152
False negative, see Type I error
False positive, see Type II error
Fisher's exact test, 69, 72
Forecast, 40, 208, 226
Found data, 16
F-ratio, 68
Fraud, 21
Frequency plot, 161

Geometric mean, 125
General Estimating Equations
  (GEE), 217ff
Generalized Linear Models (GLM), 217ff
Goodness of fit, 177, 188, 229
Graphics, 143, 149–173
  for large data sets, 168–179
  grammar, 172
  vs p-values, 78
  vs tables, 125
Grid
  line, 150
  points, 208

Groundwater, 209–211
Grouping results, 29
Group-randomized trials, 82
Growth, 125, 190, 227

Heterogeneity, 31
Hierarchical models, see HLM
Histogram, 121
HIV, 28
HLM, 217
Hodges–Lehmann estimator, 59
Hydrology, 209–211
Hypertension,
Hypothesis, 13, 234
  alternative, 13, 67, 96
  null, 16, 25, 82, 136
  ordered, 21
  post hoc, 7, 110
  primary, 67
  testable, 14
Hypothesis testing, see Tests

Immunology,
Income, 29, 125, 140, 199, 208
Independence, 40, 222
  vs. correlation, 91
Indifference region, 136
Induction, 22, 100
Interaction, 93
Interpolation, 178
Interquartile range, 196
Interval estimate, 61
Intraclass correlation, 82

Jackknife, 229
Jonckheere–Terpstra statistic, 91

Kepler's Law, 5, 112
k-fold resampling, 228
Kinetic molecular theory, 179
Kruskal's gamma, 131
Kruskall–Wallace test, 95
k-sample analysis, 89, 112

Labels, 155
Least absolute deviation, 90
Least squares, 185
Legal applications, 40, 69, 105–107, 149,
  184, 195

Legend, 152
Leukemia, 107
Likelihood ratio, 18
Likert scale, 158
Linear vs. nonlinear, 178
Link function, 217
Litter, 40
Location parameter, see Mean,
  Median
Logistic regression, 206, 209, 213
Logistic growth, 190
Losses, 22, 60, 110, 113
  absolute deviation, 90
  jump, 58
  monotone, 58
  square deviation, 60
  step function, 58

Mail, 41
Main effect,
Malmquist bias, 134
Mann–Whitney test, see Wilcoxon
Marginals, 19, 28, 69, 73, 125
Maximum, 53
Maximum likelihood, 57, 61
Maximum tolerable dose, 15
Mean absolute deviation, 196
Mean, arithmetic, 147
Measurement device, 30
Measurements, 123–125
Median, 130
Medical device, 120
Medicine, 206
Meta-analysis, 110–111, 219
Minimum, 53
Minimum effective dose, 24
Missing data, 38, 54, 98, 102, 120
Model
  building, 214
  choosing, 178
  fixed effect, 214
  nonlinear, 178, 202
  nonunique, 179–181
  range of application, 178
  general linear, see GLM
  parametric vs. nonparametric, 234
  physical, 191
  selection, 178
Monotone function, 58, 89, 178

MRPP, 186
Multifactor designs, 93
Multiple
    endpoints, 38
    tests, 142
Multidimensional displays, 167
Multivariate analysis, 77, 142, 230
Mutually exclusive, 14

Negative binomial, 218
Negative
    contagion, 218
    findings, 111, 134, 136
Neural network, 211
Neurology, 134
Newton's Law, 21
Neyman–Pearson theory, 17
Nominative label, 156
Nonlinear, 163, 178, 202, 222
Nonresponders, 39, 140
Nonsignificant results, 111, 134, 136
Nonnormal alternatives, 77
Normal approximation, 71
Normal distribution, 186
Normal scores, 77
Nuisance parameters, 222
Null hypothesis, 136

Objectives, 4, 27, 52, 58, 99, 110,
    182, 226
Oblique rotation, 209
Observational studies, 110
Observations
    dependent, 81
    exchangeable, 65, 70, 112
    identically distributed, 70
    independent, 6, 40, 205
    transformed, 82, 90
Odds ratio, 145
One-sided vs. two-sided, 19, 37, 69
Oral presentations, 171
Ordinal data, 122, 125
Outliers, 55, 120
Overfitting, 214

Paired observations, 79
Panel estimation, 228
Parameters, 33
    location, 33

scale, 33
shift, 59
Paranormal, 8
Paternity, 104–106
Patterns, 7, 28, 45
Percentages, 145
Percentiles, 103
Permutation test, 16, 75, 77, 85, 101,
    112, 186
Perspective plot, 159
Phase III trials, 24
Physician, 38, 120, 214
Pie chart, 161
Pivotal quantity, 65
Placebo, 44
Plots
    contour, 160
    four-plot, 54–55
    histogram, 54
    lag-plot, 54
    normal Q-Q, 54
    3-D, 159
    time plot, 54
Plotting region, 153–155
Plotting symbol, 156
Poker, 10
Polar coordinates, 161
Polyline, 163
Polynomial, 180, 205
Population, 5, 10
Population statistics, 5
Poverty, 182
Power
    Curve, 34
    post hoc, 135
    related to effect size, 96
    related to sample size, 32
    reporting, 135
Precision, 127
Precision vs dispersion, 127
Prediction, 182
    error, 229–230
Principal components, 209
Pricing, 231
Proc ARIMA, 186
Proc Means, 51
Proc Mixed, 83
Proc t-test, 196
Protocol, 8, 15, 32, 57, 137

Proxy variable, 183
p-value, 10, 99, 130, 135
   vs association, 131

Quality assurance, 53
Quantile regression, 199

Radioimmune assay, 75, 178
Random number, 7, 128
Randomized response, 55
Randomizing, 42–43, 118
Ranks, 71
Rare events, 9, 123
Rates, 145
Ratio, 218, 225
Raw data, 51, 114, 142
Regression
   coefficients, 185
   Deming (EIV), 76, 193, 196–198
   LAD, 193–196
   linear vs. nonlinear, 185, 194, 202
   methods, 186, 190
   multivariable, 205–215
   negative binomial, 218
   non-parametric, 191
   OLS, 177–190
   Poisson, 218, 221
   quantile, 190–200
   scope, 178
   stepwise, 205, 225
   stratified, 183,201
Regulatory agency, 18, 34, 40
Rejection region, 18, 132
Relativity, 22
Repeated measures, 81, 214
Report, 209
Resampling, 228
Residuals, 126, 235
Response variables, 28
Risk-benefit analysis, 114
R-squared, 205, 230
Rugplot, 130

Sales, 42, 227
Sales pitch, 30
Sample, 6
   Cost, 30
   representative, 10, 140
   sequential, 36–38

   size, 32–34, 37, 71, 84, 121, 140,
      198, 211
   splitting, 225, 227
   stratified, 140
   universe, 144
Scale alternative, 81
Scale parameter, 32, 34
Scatterplot, 168–170
Seismology, 134
Semiparametric methods, 102
Sequential trials, 38
Serial correlation, 54
Shading, 167
Significance
   causal, 185
   practical, 188
   statistical, 98
Significance level, 130
Significance level vs. *p*-value, 235
Significant figures, see Decimal
   Places
Silicon implants, 44
Simpson's paradox, 184
Simulations, 80, 91, 112, 128
Slope, 28, 76, 171, 180–182, 187,
   197–198, 206
Software, 33
Stability, 231
Standard error, 127
Stationarity, 188, 231
Statistic,
   Durbin-Wu–Hausman, 222
   Gehan-Breslow, 75
   Hotelling's T^2, 77
   Jonckheere–Terpstra, 90
   Mantel–Cox, 75
   Tarone–Ware, 75
Stein's paradox, 65
Stepwise regression, 205, 225
Stochastic, 233
Strata, 183
Studentize, 63
Subgroups, 15
Subjective
   data, 91
   probabilities, 108
Sufficient statistic, 65
Surgery, 10
Surrogates, 28, 183, 187

Surveys, 7, 30, 120
Survival analysis, 20, 73, 75

Tables vs graphs, 125, 158
Terminology, 233
Tests
    analysis of variance, 93
    assumptions, see Assumptions
    bootstrap, 80–81
    chi-square, 72–73
    exact, 171
    Fisher's exact, 18, 72
    for equality of variances, 85
    for equivalence, 79
    for independence, 73
    F-test, 89
    Kaplan–Meier, 75
    Kruskall–Wallace, 95
    k-sample, 89, 112
    locally most powerful, 81
    MannWhitney, see Wilcoxon
    matched pairs, 142
    most powerful, 96
    multiple, 97, 133
    one-vs. two-tailed, 19, 37, 69
    O'Brien's, 80
    permutation, 16, 75, 77, 85, 101,
        112, 186
    rank, 113
    reporting, 141
    Smirnov, 81
    t test, 68, 76
    types, 71
Texture, 167
Time series, 134, 225
Time-to-event data, 73
Title, 166
Toxicology, 41
Training set, 213
Transformations, 80
Treatment allocation, 45, 118
t-test, 76

Type I and II errors, 16, 30, 33, 71, 235
Type I and Type II censoring, 75
Type II error vs. power, 33, 235

Unbalanced vs balanced design, 93,
    95–96
U-statistic, 71

Vaccine, 36, 72
Validation, 225ff
    need for, 209
Variable
    categorical, 72, 122
    confounding, 181, 207
    continuous, 34, 61, 123, 127
    exchangeable, 86
    explanatory, 206, 211, 218, 227–230
    indicator, 189
    predictor, see explanatory
    primary, 28
    proxy, 183
    response, 28
    selection of, 11
    surrogate, 28
Variance
    comparing, 85
    estimator, 230
    function, 218
    inflation factor, 82
    ratio, 80, 85, see F-Test
    sandwich estimate, 222
    unequal, 79
    verification, 226
Variation, 5
Viewpoint (for graph), 159
Virus, 119, 125, 183
Voting, electronic, 208

War, 100
Weather, 18, 38, 208
Welfare, 190, 199
Withdrawals, see Dropouts